PRAISE FOR *OUR SHARED STORM*

"Hudson's innovative and exciting publication is, simultaneously, a consideration of the relationship between climate fiction and climate policy, a highly readable and teachable set of climate stories, and a critical intervention in what climate fiction is capable of achieving in the 'real' world."

Adeline Johns-Putra, Professor of Literature,
Xi'an Jiaotong-Liverpool University, China

"Andrew Dana Hudson's *Our Shared Storm* is a fascinating thought-experiment in imagining worlds to come. Through a set of common characters kaleidoscopically revealed, readers are granted perspectival narrative access to a skein of political, cultural, and philosophical views that, along with their attendant actions, will shape the planet for worse—or, perhaps, better."

Christopher Schaberg, author of
Searching for the Anthropocene

T0044554

Our Shared Storm

OUR SHARED
STORM

A Novel of Five Climate Futures

Andrew Dana Hudson

FORDHAM UNIVERSITY PRESS   NEW YORK   2022

Fordham University Press has no responsibility for the
persistence or accuracy of URLs for external or third-
party Internet websites referred to in this publication
and does not guarantee that any content on such
websites is, or will remain, accurate or appropriate.

Fordham University Press also publishes its books
in a variety of electronic formats. Some content that
appears in print may not be available in electronic
books.

Visit us online at www.fordhampress.com.

Library of Congress Cataloging-in-Publication Data
available online at https://catalog.loc.gov.

Printed in the United States of America
24 23 22 5 4 3 2 1
First edition

CONTENTS

INTRODUCTION

One Story, Five Worlds

This book is about the future of our climate. Given that our summers now regularly feature Arctic heatwaves and wildfire blood skies, our winters reach with polar vortex fingers all the way down to Texas, and hundred-year storms hit every couple of months or so, it may seem that catastrophe is a done deal. As grim as things are, however, I believe we still have options. Depending on the choices we make now and in the next few decades, we can either steer into or swerve away from the worst of the damage still to come. We can fall into disarray and civilizational ruin—definitely an option! Or we can reduce our emissions to mitigate global warming. We can shore up our cities, agriculture, and infrastructure to adapt to rising seas and heavy weather. And if we make very good choices indeed, we may even stabilize the planet, clean up our carbon waste, and gently, gently walk back from the edge of disaster.

This book is also a work of futuristic speculative fiction. It's an unusual one, however, in that it explores not just one potential future, but five. Each of the five stories that follows is set in the same place and time: Buenos Aires in the year 2054, during the annual global climate negotiations called the Conference of the Parties, a.k.a. the COP. The stories feature an overlapping set of characters, including four characters who appear in all five stories. Depending on which future the story takes place in, however, events unfold differently. The characters are different people, having lived, for the thirty-odd years between now and then, diverging lives. And, depending on the future, the COP will be a very different kind of gathering as well.

Even stranger, these five futures are not entirely my own whimsy, but are inspired by a set of climate modeling scenarios called the Shared Socioeconomic Pathways (SSPs). These scenarios were devised to inform the Sixth Assessment Report of the Intergovernmental Panel on Climate Change (IPCC). (At time of writing this report is still being prepared.) The SSPs and this book both use a futurist technique called "scenarios thinking." Rather than trying to predict how history will unfold—picking a single future out of many unpredictable and contingent branching paths—one instead creates a set of future visions that represent major trends or counterposed possibilities.

In the case of the Shared Socioeconomic Pathways, the five scenarios are plotted on a chart where the axes are "challenges to mitigation" and "challenges to adaptation." This chart is divided into four quadrants, with additional space carved out in the center for futures with moderate challenges to both mitigation and adaptation (see fig. 1). These scenarios are connected to specific quantitative projections of various metrics of human civilization (population, GDP, land use, emissions of different gases, and so on). However, undergirding the pathways is also a set of descriptive narratives—"the roads ahead"—originally proposed by an international group of scholars (O'Neill et al.) in 2017, which I will summarize here. Feel free to bookmark these pages to refer back to, if you find yourself interested in just what assumptions underlie each story.

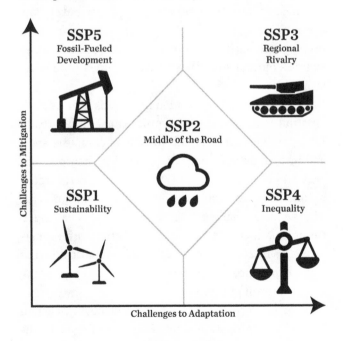

Figure 1

SSP1 is the sustainability pathway—"taking the green road," to use the catchphrase developed by the original narrative authors. Challenges to both mitigation and adaptation are low, and so the world makes a relatively smooth transition to more sustainable lifestyles and economies focused on human well-being. Inequality decreases. This is considered the most optimistic scenario, the "best" future to shoot for with the greatest chance of keeping global warming under 1.5 degrees Celsius (1.5°C).

SSP2 is a "middle of the road" scenario in which present trends largely continue. Some slow progress is made, but there are also challenges. Development and climate action proceed unevenly. In this scenario, circumstances do not necessarily stay the same, but no single trend swerves our current trajectories into one of the other scenarios.

SSP3 is a scenario in which both mitigation and adaptation challenges are high, producing a "rocky road." Little is done internationally to address climate change, and instead nations focus on preserving their own energy and resource security. The world sees resurgent nationalism, regional rivalries, and military conflicts. The population balloons. Because of the high challenges, this is usually considered normatively the worst scenario, but it's interesting to note that this scenario does not actually produce the highest levels of emissions, owing to slower economic growth.

SSP4 is "a road divided," defined by inequality. Though progress on mitigation is possible, including some investment in renewables alongside some continued use of coal and oil, the high challenges to adaptation mean that poor countries and poor people experience much worse impacts from climate change than do their wealthy counterparts. Inequality is felt in a diverging economy as well. The widening gap between haves and have-nots leads to frequent unrest.

SSP5 is called "taking the highway." It describes a world where little is done to curb emissions, and instead fossil fuels are burned to provide cheap energy for powering economic growth and development. The world is increasingly globalized, with significant investments in health and education to improve the lots of low-income populations. Climate change is treated as an adaptation problem focused on local environmental degradation, which gets handled without slowing overall global emissions. The world looks to technology and geoengineering to solve the climate crisis.

The first sections of the Sixth Assessment Report, published in 2021, detailing the physical science findings on climate change, feature a slightly different formulation of the SSPs, skipping SSP4 and instead offering two different versions of SSP1, one in which warming is kept under 1.5°C and another in which the planet stabilizes at slightly under 2°C warming. SSP4 was omitted because, in terms of pure climate outcomes, it didn't offer a pathway significantly different from that of SSP2 (IPCC 2021). I should also note that some scholars have argued that SSP5 is increasingly unlikely because coal use is already declining (Burgess et al. 2020). The burning of coal at an ever-accelerating rate was one of the assumptions on which the gargantuan emissions of that scenario were predicated. Nonetheless, I find both SSP4 and SSP5 to be instructive scenarios to consider and have included them in this book.

When I first read these narratives, I was struck by how much they felt like speculative fiction stories—specifically *climate fiction* stories, the framing of which I discuss in an afterword to the book. Indeed, one group of scholars (Nikoleris, Stripple, and Tenngart 2017) has already critiqued the SSPs by connecting them to the literary visions of climate change in popular novels. I had also seen speculative fiction that used variations of scenarios thinking, including Kim Stanley Robinson's classic *Three Californias* trilogy, published between 1984 and 1990. Tobias S. Buckell's story "A World to Die For" (2018) features timeline hoppers who travel between worlds that represent another set of IPCC scenarios, the Representative Concentrated Pathways (RCPs). However, no one to my knowledge had yet attempted to turn the SSP narratives directly into fiction.

I decided to write a set of stories based on the Shared Socioeconomic Pathways—both to bring the IPCC's work to a broader audience and to use speculative fiction's unique toolbox to deepen our collective understanding of these scenarios. Fictional illustrations could further develop the SSPs as communication tools to help individuals, institutions, and policymakers see how their choices and investments push us toward different possible futures. Those trying to understand the SSPs could read these stories to get a sense of how these grand global narratives might map onto their own communities, what being in one SSP future versus another might mean for the texture of their lives. In short: What do these futures look, smell, taste, sound, and feel like? And, through the writing process, I could look for new cracks, contradictions, or insights in the SSPs by squeezing them through the creative wringer.

To flesh out this idea, I visited the International Institute for Applied Systems Analysis (IIASA) in Vienna. IIASA hosts a database of quantitative projections that scholars have attached to the SSPs, allowing one to check, say, the hypothetical GDP of China in 2050 in SSP1, or chart Brazil's population changes over the coming century in SSP3, or compare global emissions trajectories of various greenhouse gases in each SSP. The researchers at IIASA I spoke to were intrigued by the idea of using fiction to illustrate the SSPs, but they posed a fundamental question: How could I make sure that the illustrative details of each story could be clearly attributed to that story's SSP, and not, say, to the place and time of the setting or the idiosyncrasies of different character dynamics? After all, illustrating the SSPs means highlighting their differences, writing about them not just as themes grounding individual stories but as alternate possible future histories.

I resolved that instead of writing an independent story set in each SSP, I would write one story five different ways. It would be a kind of literary thought experiment, eliminating as many variables as I could. Using the same city and the same core characters, I would explore how each pathway shapes who they are and how they handle the events of the story.

All stories would also be set in the same year, in the middle of the twenty-first century—not so far out that

I couldn't reasonably give the stories a shared climate event to circle around. This would put them in the heart of the challenges and opportunities imagined by each narrative, when the investments in adaptation and mitigation being promised today will have been either successfully completed or kicked down the road. Which way those investments go will likely push the world into one or another SSP by the 2050s, but the full consequences of those choices would remain a generation off.

And because I wanted my stories to have a global scope, just like the SSPs, I chose to set them at the annual Conference of the Parties and to focus on the future of the ongoing climate negotiations conducted under the auspices of the UNFCCC—the United Nations Framework Convention on Climate Change.

A few words about the COP, before we get to the stories. As research for this book, I attended COP24 in Katowice, Poland, in 2018. The location was at once earnest and ironic. Katowice is in the heart of Polish coal country, and indeed the Katowice pavilion was a literal shrine to coal and its aesthetics, displaying coal jewelry and coal-based cosmetics. There was supposed to be, buried in there, a narrative about "black to green" economic transformations. But for many in the activist circles I ran with that week, the coal was a defiant signal of intransigence by the Polish government that presided over the conference. The winter air outside the conference venue tasted of stale exhaust, and throughout the week I heard people complain about the smoggy cold: "the smold."

I attended as part of Arizona State University's observer delegation. ASU, like many of the world's academic institutions with an interest in climate, was granted a handful of badges to send students and faculty each year. While some institutions sent coordinated groups, I was left to my own devices, only occasionally bumping into the one other ASU student attending week one.

As an observer I primarily rubbed elbows with other yellow-badged observers: activists and researchers, mostly, folks who filled out the civil society "constituencies." The constituencies were grouped into ENGOs (Environmental Nongovernmental Organizations), RINGOs (Research and Independent NGOs), YUNGOs (Youth NGOs), BINGOs (Business and Industry NGOs), TUNGOs (Trade Union NGOs), IPOs (Indigenous Peoples Organizations), farmers, WGC (the Women and Gender Constituency), and LGMAs (Local Governments and Municipal Authorities). "Nongovernmental" was the operative word for all these constituencies, for only actual nation-states got to have an official say in the negotiations. The rest of us were there merely to advise, agitate and, obviously, observe.

The other side of the coin, aside from United Nations staff, were the pink-badged "parties," individuals who served on delegations sent by the various nations that were parties to the UNFCCC. Some of these individuals—such as celebrities or young people—were there more to represent their nation in an abstract sense, as opposed to actually participating in the negotiations. Some ran side events or worked the booths and pavilions that most countries set up to showcase their climate progress, their ideas, and national character. Many national delegations were a mix of ceremonial representatives and wonky diplomatic operatives.

I was there technically as a RINGO, but I had found housing with a group of largely American youth activists called Care About Climate. Care About Climate's signature campaign was trying to popularize the "climate sign," that is, holding one's hand up in a C-shape, analogous to the 1960s' peace fingers. They also had the connections to get their hands on some half-dozen badges each year. I met part of the Care delegation in the echoey, brutalist hostel they had arranged, along with other activist groups, a few kilometers from central Katowice. A few of them had been to the COP a couple of times previously, and so I was glad when they took me under their wing. They introduced me to other YUNGOs and let me shadow them into activist gatherings I might otherwise be excluded from, such as daily meetings of the Climate Action Network (CAN, a large network of environmental groups) and the notorious first Saturday CAN Party. From this social seed I was able to schmooze and make friends, and within a day or two I had a small posse of fellow travelers I recognized in the hallways, with whom I could hang out, eat, and share tips, and who could help me make sense of the COP's colossal diplomatic and bureaucratic apparatus.

The COP takes place for two weeks once each year, in a different country each time, but this conference is part of an ongoing diplomatic process that includes midyear meetings in Bonn, Germany, and that has been snowballing in complexity and ambition since the UN-FCCC began in 1992. The UNFCCC is a treaty establishing the framework for developing a more robust treaty on climate change. The actual treaties signed are colloquially called the Kyoto Protocol and then the Paris Agreement—named after the cities that hosted the COPs where these agreements were made. Therefore, some years the COP is a more important diplomatic event than others, with world leaders showing up to try to strike a grand bargain. Notoriously, Copenhagen, in 2009, was intended to be one such year, but the negotiations fell through. In "off years," such as 2018 in Katowice, the parties still gather to report progress and opinions, massage the details of past or future agreements, and urge each other to greater ambition. The COP also serves as a grand convention on all things climate change, gathering tens of thousands of people to pick at every facet of one of the greatest challenges human beings have ever faced.

The plan to solve that challenge is to get the world—including the major petrostates like Russia, Saudi Arabia, and the USA—to sign on, in consensus, to a legally binding but largely unenforceable agreement to cut emissions and pledge money toward fighting climate change. Parties will be held accountable by the peer pressure of the international community. Over time every country is expected to ramp up its commitments—since reaching an agreement that actually promises to do what scientists say is necessary to keep the world at 2°C warming has so far proved untenable.

For a long time, the moral and political quandaries at the heart of these negotiations swirled around issues of who was being asked to make which sacrifices. Should poor countries have to curtail the fossil-fueled development rich countries have already enjoyed? Should rich countries bribe poor countries to forgo future emissions? Who pays for the damage to sinking island nations that had very little to do with driving the rising of the seas? This seems to me like a counterproductive way to approach solving such an important collective problem—especially when switching to renewable energy and cutting down on pollution now seem to be excellent investments. But whether because of the structure of the negotiations, the nature of the negotiating parties, the mood at the COP, capitalism, neoliberalism, the lack of narratives, public apathy, or pure human pettiness, these have been the central questions of climate politics for the last thirty years.

The stories in this volume do not try to answer those questions, but instead ask: What might the central questions in climate politics be in another thirtyish years, at COP60? What has been accomplished by that point in each of the five SSP futures? What is still being debated? Who are the power players in each scenario? What new issues become contested, and what are the terms of those contests?

Too much more detail and commentary, and I risk sprawling this introduction into the realm of spoilers. Instead, I'll offer just one last bit of guidance before asking readers to plunge in. Each story stands alone, but they are intended to be connected, forming what the publishing world calls a "fix-up novel." The stories have been sequenced not from SSP1 to SSP5, but in an order I hope will create an emotional arc, a sort of metanarrative. In the end, however, much in these stories is left up to the reader's interpretation.

Our Shared Storm

POLITICS IS PERSONAL

First Monday

The opening ceremony had birds. Pale, brown blurs, arcing over the plenary in elaborate formation.

"Dear friends, we are in trouble," the outgoing COP president said from the podium. It was a traditional line, Noah knew, and who got to say it during the opening ceremony was a matter of some diplomatic horsetrading. "Despite our sixty years of effort, the world stands poised to cross 550 ppm, locking us in for more than 2.5 degrees of *catastrophic* warming."

Were birds a new flourish? Noah watched them whirl, thinking back. COP59, Singapore, no ceremonial birds. COP58, Cape Town, no birds. COP57, New York, definitely no birds—except the ones outside, the pigeons that had accosted him on Roosevelt Island.

1

"We *must* come out of this COP with a real *and* realistic London rulebook for mitigation and adaptation. And we *must* each *commit* to raising sequestration ambition, so that we can *finally* do the right thing: put it *back* in the ground."

Scattered applause. The outbound Singaporean COP president handed a tiny gavel to the new Argentinian COP president, who clacked it on the podium, officially opening the Sixtieth Conference of the Parties to the United Nations Framework Convention on Climate Change.

Noah had heard it all before, the exhortations to commitment and ambition, with stresses on specific words to help news algorithms parse the subtext. Once he had fiercely taken notes, hoping to guess, from that subtext, the character of the negotiations to come. But this year he just zoned out, followed the aerial dance of the still-swooping birds. What kept them coordinated, he wondered. Tiny insect drones to chase? Or chips in their brains? He gestured at his program, but the screen just popped up a page on *Furnarius rufus*—the rufous hornero. Apparently these were Argentina's national bird.

The COP60 president began a somewhat more conciliatory speech, filled with lines about "common but differentiated responsibilities" and "operationalizing the win-win solutions"—just in case any parties had gotten the impression that *they* were being asked to increase ambition. Buenos Aires had been host to COP10, and the Argentinians were keen to make that half-century anniversary a self-congratulatory one.

Noah's phone buzzed with priority. It was Marta Tolmbly, US head of delegation: his boss.

"I'm getting sucked into last-minute DC scrambles," the message said. "Can you pinch-hit for me in these constituency sit-downs? Details attached."

Noah scanned the meeting list and silently groaned. Marta's assistant had already updated his schedule, bumping him out of several negotiation sessions. He wondered if he could reschedule with the constituencies one by one, but that would mean missing out on vital COP activities, like beating the rush at the lunch buffet and flirting with the barista at the German pavilion. At least the first sit-down was coming up soon, an excuse to bail early on the opening ceremony.

He half-stood, shuffled to the plenary hall exit. He looked up one last time, marveling as the tan flock seemed to mime a crashing wave. A fleck of white shit landed on his pink badge.

First Tuesday

Noah hustled through the passageways of Spaceship COP. The event was held this year on the sprawling convention campus that had, a promotional video told him, replaced Buenos Aires's aging naval industrial complex on the banks of the Río de la Plata. Probably a nice place for a summer stroll—marshy ecological preserve on one side, polished vintage ships propped up for examination on the other. But any desire by the Argentinians to actually show off this post-postindustrial playground was overpowered by the need to keep the ninety thousand easily confused inhabitants of the COP-cosmopolis contained and securified. The result: a maze of snap-together tent tunnels, modular meeting rooms, and prefabbed bathrooms. Noah tried to remember if he'd caught a whiff of sea breeze on the brief walk from his hotel shuttle to the indoor security theater.

It was ironic, Noah thought, not for the first time, that the COP always fastidiously cut itself off from the very thing it was trying to save—the climate. And not very well, either; rooms and passageways were either frigid with blasting air-conditioning or baked like a lunchbox left out in the sun. Sometimes both in the same session.

Noah had a long list of such gripes, but he kept coming back, year after year. He liked to think that was what made him good at his job: his intimate familiarity with the mediocrity at the heart of all large human endeavors. History would mark their efforts as either a great triumph or—to be realistic—a dismal failure. But doing the work each day was all sweaty palms and papercuts, staving off hanger and sleep deprivation long enough to productively fiddle with a comma on a document that didn't matter. The better one understood that and learned to navigate the mundane annoyances, Noah believed, the better one could actually get something done.

So it was a sour but self-satisfied mood that carried him across the venue to the cubicle honeycomb of Small Meeting Room Cluster G. Along with dumping these constituency sit-downs on him, Marta had been shifting the negotiation roster like crazy, and somehow he kept getting swapped out—even off parts of the text he'd worked hard on. This, his fourth civil society meeting in two days, was the most exciting thing left on his agenda.

He checked his phone: two minutes late. He scrolled through his messages for another ninety seconds before going in.

"Sorry I'm late," he said. "And sorry I'm not Ms. Tolmbly. Noah Campbell. I'm one of the US negotiators and kind of a jack-of-all-trades in our delegation, so Marta asked me to represent her today, as she was unavoidably detained in Washington."

The woman who shook his hand had yellow hair and a yellow-banded observer badge. With the practiced eye of a thirty-seven-year-old single man, Noah checked for a ring. There wasn't one, though he wasn't sure what courtship capitalism was like in Scandinavia these days. She was attractive, he thought, even by the standards of the glamorous European women who so often filled the halls of international diplomacy. But there was something about her taut-angled cheeks and the drape of her spartan clothes that made him think she didn't come from that private-school-to-NGO pipeline. In a burst of occidentalist imagination, Noah visualized the woman as an armored Valkyrie, singing from a mountaintop.

"Saga Lindgren," she said. "I understand. I was surprised she agreed to make time for me in the first place."

"We always get overly ambitious scheduling the first two days," Noah said. "But we think these one-on-ones with the constituencies are a good way to set expectations. On both sides."

"Well, you've certainly done a good job so far," Saga said, and Noah wondered how much annoyance was buried under her chipper Swedish accent. Maybe it was just her jet lag.

He glanced at the meeting info on his phone. Saga represented Give Us Repair & Reparations!—GURR!—one of the ever-shifting coalitions that fought like snakes for factional dominance within the Environmental NGO constituency. Unlike the studiously independent RINGO researchers and the YOUNGO youths who tempered their radicalism with school-field-trip deference, the ENGO groups had real political demands. Better cut to the chase.

"Indeed. Well, what can the United States do for GURR! this COP, Miss Lindgren? I don't have much power, but I am skilled at getting yelled at."

"Did you listen to the speeches at the opening ceremony yesterday?" Saga asked, not correcting him to "Ms." or "Mrs."

"Mostly. Some of them. That thing with the birds was incredible."

"Perhaps you noticed, then, that there was no mention whatsoever of loss and damage."

Noah shifted uncomfortably. He hadn't noticed, but he hadn't needed to. The US had burned a lot of political capital in the conference planning process cutting loss and damage language from those speeches. Not surprising that someone had gotten ticked off enough to talk shit to activists like GURR!.

"As you know, the London Affirmations we've all signed on to very plainly lay out expectations for L&D," Noah said carefully. "Given that we are here to fill out the London rulebook, it's not surprising that the speeches focused on the trickier, as-yet-undetailed facets of that agreement."

"Ah, no reason to discuss the notoriously untricky, uncontentious issue of loss and damage, yes?"

He didn't need his tone-parsing app to recognize the sarcasm slathered over her words.

"Okay, loss and damage is contentious. No one is arguing otherwise," Noah said. "But it's not as . . . up for grabs in this particular COP, this rulebook. The rules on adaptation finance and sequestration accounting are where the real meat of the negotiations will be. Can you blame us for wanting parties to keep their eye on the ball?"

Noah winced, realizing too late that this was a tacit admission that the US—the "us" in his sentence—had pushed loss and damage out of the opening ceremony discourse. He should have had more caffeine today. But Saga didn't seem interested in that particular gotcha. Instead she glanced down at her phone, checking, Noah assumed, for the meaning behind his "eye on the ball" Americanism.

"So you have them take their eyes off the baseball of loss and damage, instead?" she said.

"That's not quite how baseball works."

"Mr. Campbell," she pronounced it "Camp-Bell," "every year the sums pile higher. Pakistan, Kenya, Mexico, Haiti, Manila, the Killed Islands. Whole cultures are being ripped apart by heat and floods and storms that weren't supposed to happen. I myself lived through Thor's Year as a youth. I have seen how, decades later, the communities those storms hit are still struggling to find stability. These people deserve reparations from the Historically Polluting Parties. At GURR! we do not think it is a distraction to talk about the victims of climate change at the climate change negotiations, no matter the task at hand. Not while refugees throw themselves into the Mediterranean and wildfires burn down your own California."

"I'm from California, okay? Why do you think I'm here?" Noah said, loud and hot, surprised at his own vigor. For some reason Saga's needling was making him lose his cool. Perhaps it was compounding the frustration he felt at Marta dumping these piss-taking meetings on him, sidelining him from negotiations. Or perhaps, a voice in the back of his head noted, it was because the needling came from a beautiful woman, and he had spent the year burning out on one pop-up dating app after another.

Usually Noah didn't take such callouts personally. He was used to speaking for institutions that weren't as pristine or well-behaved as anyone would have liked, himself included. But the alternative was walking out, refusing to serve, and that just meant ceding his power to people who cared about the climate less than he did.

He tried to get back on track. "You know that some parties will happily swerve discussion into, uh, intractable territory to kill time in sessions they don't want to see make progress."

"Of course," Saga said. "I've seen American allies like Saudi Arabia do it many times."

"I was thinking more like China, but sure. Point is, we want this process to work. We want a complete rulebook. I guarantee you that's Tolmbly's number one priority. So we are going to do what we can to hit that marker, so the process can move forward. Kyoto, Paris, Glasgow, London—we can't keep going back to the drawing board. The US is all in on the London Affirmations."

"London demands that loss and damage be entwined into every facet of climate action and management," Saga pressed, not giving him an inch, even after his Very Strong Statement of Commitment.

"Yeah, London was a big win for you guys. GURR! totally deserves to own the L&D ambition in that agreement. But look, loss and damage isn't going away. What *is* going away is our precious, precious time to get our energy-to-emission ratios high enough to start doing serious drawdown. I don't like it either, but it's going to be all loss and all damage, forever, unless we push the boulder of energy transition up that hill."

Saga wrinkled her nose. "Is that another baseball reference? Do Americans still play that sport?"

"No," Noah laughed. "It's Greek, I guess. Sisyphus."

"Oh, of course. The lying, greedy merchant king who almost ruined the whole system by trying to live forever. I see why you think of him. Very American."

Noah opened his mouth to protest the burn, but then Saga laughed too, a resonant peal. She leaned back in her chair and smiled at him—thinly, but not without warmth.

"So, are you being honest?" she asked, still smiling. "Those are the expectations we should have for the US this COP? You'll be too focused on holding your big rock to spare a glance down the mountain at the victims of American predatory delay?"

"Not exactly how our PR people would have me put it," Noah sighed. "But yes. L&D is not a discussion priority for us this COP. That's the directive from Marta, from the State Department, from the White House. You can push us on lots of other issues, but I honestly think it would be a waste of energy to try to get us on this. I wish I had better news for you. But I will say I think you'll be pleased with where we land on the repair half of GURR!'s mission. Lots of talk stateside about the home run opportunities to lead innovation in the 'cleanup economy.'"

"Then this was a productive meeting after all," Saga said.

She got up and grinned again in that thin, chipper Nordic way. She shook his hand and walked out of the cubicle, leaving Noah feeling deflated that the exchange had gone so poorly and elated that it had ended on something so close to a high note.

The tête-à-tête had been strangely intimate, at least compared to the abstracted, sometimes alienating experience of intoning the desires of a nation in a bureaucratic ritual that had been unfolding for sixty years. Saga had talked to him like a person, an individual. He doubted she had intended this, he was sure he was just getting high on the attention of a pretty woman at a moment when his career doubts were mounting, but still—he felt seen.

First Wednesday

"Did you read ECO this morning?" Diya asked over lunch the next day.

"Of course," Noah lied. "Why? What's in it?"

Usually he did at least scan the activist news pamphlet that got handed out every morning of every COP, but today he'd shown up early, and the Climate Action Network vending bots had still been empty when he'd dashed by to make a core delegation huddle. Marta had bumped up the meeting without telling him; he had to hear about it from the personal assistant of a famous gamer-activist joining the delegation next week.

"Someone on your team screwed up," Diya said.

The Indian academic, a COP friend for several years, handed him her copy of ECO, open to a quarter-page section headlined: FUMBLING SISYPHUS'S BASEBALL.

"Dammit," Noah said, and he read the thinly anonymized report of his conversation with Saga the previous day. "Dammit," he repeated occasionally.

For all the US loves mixing metaphors, the article mocked, *they seem uncomfortable mixing the writing of accounting rules for a still-hypothetical sequestration system with real progress addressing the loss and damage already on the books—mixing their gallant saving of the future with consideration for the millions of present-day climate victims asking for justice. As always, the American melting pot prefers to boil away inconvenient history, so they can feast on the hearty stew of the next opportunity.*

Diya was watching him, amused. He knew she knew that he was the unnamed US delegate being skewered by the ECO piece. She reached out and patted his hand sympathetically.

"Whatever metaphors I mixed couldn't have been as bad as this garbage," Noah complained. "I mean, does a melting pot even make stew?"

"You have fine metaphors, darling," Diya said. "Though I suspect the substance of this piece will be more concerning to your delegation than the style."

Diya Kapoor was the prodigal scion of a rich filmi clan in Mumbai. Her temperament had steered her away from the family business and into the grim scientific field of cataloguing and modeling Earth's descent into the hothouse. They'd met at Noah's first COP, six years earlier in Havana. It had been a mess of an event, and when negotiations had broken down he'd morosely hopped from club to club until he found one that wasn't packed with other miserable delegates. Diya was there too, having fled the blaring Bollywood pop mashups that were having a wave year in tourist-facing Spanish-language discos. They shared mojitos, and Diya had methodically walked him through the atmospheric physics of climate change, explaining just how little the actual text of the climate agreements could do to swerve the planetary forces the industrial age had set in motion. And then, just as methodically, she had taken him back to her rented casa and screwed his brains out. They'd been flirty friends ever since.

"I'm off my game this year," Noah said. "Don't even have jet lag, and still I've been sleepwalking through this whole event. I think Marta is boxing me out of the negotiations. She'll probably use this to kick me off the delegation for good."

"Are you trying to seduce me?" Diya chided. "You know I have a weakness for self-pitying men."

Noah looked up from his plate of pasta, curious to see if the joke would stir his libido. Diya was seventeen years his senior, but that had never stopped them before. Still, he was too frustrated to fantasize. He stabbed at the gnocchi with his wooden fork.

"Anything quite so exciting for you this COP?" Noah asked.

"Just Helene, dancing around the South Atlantic," Diya said. "Don't tell anyone, but it's the best part of my job. I remember when there used to be a hurricane *season*, they were so predictable. These days, with so much energy in the air and water, anything can happen any time! Isn't that fabulous?"

Noah tried to pay attention, make small talk about Diya's latest meteorological obsession, the unusual pressure system building off the coast. But soon enough his mind drifted back to the pretty Swede.

"It's not like Saga is wrong here," he said, unprompted. "This 'stay on target' directive is kind of bullshit, or at least it is by the reasoning State gave. Every COP has discourse that isn't pertinent to the actual texts being debated. I mean, the fact that those discussions don't actually have much impact on the negotiations is half of what civil society complains about every year! Why bother to edit speeches and smother side events that don't change the outcome? That's a Chinese approach to dissent, not American."

"Saga? Saga Lindgren?" Diya asked. "That's who you spilled to?"

"Yeah, from GURR!. Do you know her?"

"She's feisty, knows her stuff. She showed up at RIN-GO to grill scholars about the latest damage numbers. Surprised you haven't encountered her before. She's not an activist only. She consults on L&D accounting for some parties in the Para-Developing Countries bloc. You shouldn't have gotten on her bad side, bhola." The nickname—Hindi for "one who can be easily taken for a ride"—was their running joke, but felt all too appropriate today.

"I didn't think I *was* on her bad side!" Noah protested. "She came in gunning for me, and I thought I had, you know, admirably worked to win her over."

"Well, well, well, has Noah spotted his COP-crush for the year?" Diya said, eyes mischievous.

Noah groaned. "I certainly hope not. If so, I have the worst taste and timing."

"We already knew that, bhola."

First Thursday

Noah saw Saga again the following morning, walking through the pavilions. Most countries at the COP staked out little kingdoms in the wide-rowed exhibit hall—booths or even pop-up cafés where they could promote their national achievements, seeking to win either global cachet or foreign investment. Diplomacy always involved a certain amount of peacocking. Noah wondered just how many billions of dollars of climate finance had been moved around the world by the strategic deployment of swag.

Saga was waiting in line at the Thai pavilion's juice bar. Noah speedwalked to jump into queue behind her.

"You know," he said, and Saga started. He held up his hands—*I come in peace*—and she settled that Valkyrie gaze on him. "I don't blame you for spilling what I told you—that's on me. I'm bad at op-sec or whatever. But did you have to be so mean about it? I mean, oof." He shook his hand as though burned.

"I don't do the writing for ECO," she shrugged. "But I'm sorry your feelings were hurt. Did you get in trouble?"

"Yeah, my boss really chewed me out last night."

It was true. Marta Tolmbly—still in DC ironing out her game plan with the State Department—had made him jump into a headache-inducing VR conference where she and a pack of anonymized but definitely important officials had spent twenty minutes haranguing him over the ECO debacle.

"Good," Saga said. "That means it worked."

"Just," Noah groped for words, "we're both here. We're both trying our best to save the world, however ineffectively. We probably have lots in common! I don't get why you would mess with me even though I have zero control over US loss and damage policy. What did I do to you?"

For the first time in their short acquaintance Saga looked genuinely surprised.

"Do to me? Nothing, of course."

"Do you dislike me for some reason? It's the punchable face, isn't it?"

Saga laughed and then gave him that weird smile of hers.

"I cannot dislike you. I hardly know you. None of this is personal. It's just politics, that is all."

"Politics *is* personal," Noah said. "It's about grudges and distrust and favor trading. We ally with those we want to spend time with and give the benefit of the doubt to people who laugh at our jokes."

"I did just laugh at your joke," Saga pointed out.

It was true, she had. *Well, they can't say yes if you don't ask*, Noah thought.

"Then maybe we could be friends?" he tried.

Saga shrugged again. It seemed oddly affected, as though she'd learned the gesture from a translator app.

"Why not? Let's be friends," she said. Noah was about to reply, but Saga turned around and ordered her juice. The boredly smiling, gender-ambiguous Thai bartender wai'ed and began to manipulate the elaborate, chrome juicer. Noah rocked on his heels, antsy, while Saga patiently waited on the whirring machine. When she received her puke-green drink in an origami cup, she looked at him again.

"As my new good friend," the sarcasm was back, "could you arrange for Marta Tolmbly to speak at GURR!'s 'Truth and Reckoning in Climate Accounting' side event next week? As a personal favor, of course."

"Umm . . ." Noah said. He realized the people behind him were waiting for him to order a juice. He gave the bartender an awkward none-for-me-thanks wave and stepped out of the queue. "Normally I'd be happy to, but Marta is *personally* not very fond of me right now, after the ECO piece. But look, why don't we get dinner this week, or—what do you call it—'fika,' and we can talk about how we can help each other?"

Saga sipped her juice, staring at him. She seemed to really take him in for the first time.

"Mr. Camp-Bell, do you like coming here? To the COP?" Saga asked. It was not the reply he had expected.

Around them the pavilions buzzed. Noah became uncomfortably aware of his surroundings. People of all shapes and colors shambled past, peered down at phones or papers, rubbernecked at different countries' flashing projection murals, stood clumped in conversation, jogged with purpose to their next session, or spun in circles in search of a friend. A teeming ant colony, each worker following another, blind but for the urging of pheromones, trying to build something enormous, bit by bit, even though none of them could see the whole.

"Yes," he said. "For better or worse, this is where the action is. You feel a part of something here, brilliant people trying to do the hardest thing humans have ever tried to do. It's always disappointing, but also always so impressive."

Saga nodded. "I thought you must. You arrange your life and career to get you here, yes? Then when you get here, you network and schmooze. You learn what names to drop to get on the delegation again next year. No matter the outcome, you can find your way back. So, in a way, the outcome doesn't matter, does it?"

"Hold on—" he began.

"I do not want to be here," she continued. "It is too hot outside, too cold inside. I always get sick after the COP. I found a bedbug in my hostel, and I do not like to fly. I am here because my organization asks me to be here. I come back because the work keeps being left unfinished. When I leave this Sunday, I hope I never have to return, because that would mean we have won. So, I am not here to make friends. I do not believe politics is personal. Politics begins where there are millions, not these selfish thousands here. And the millions are drowning, and burning, and starving. So, I do what I can to show those with power that *you do not do real politics here.* Not when you jostle for position and turn catastrophe into an arena of social competition. I am neither disappointed nor impressed."

"You're leaving on Sunday?" Noah blurted. Somehow, in the whole, sudden speech, it was the simple fact of her departure that floated to the top of his brain. He found himself, despite himself, trying to plan how he could spend time with her before she left. *You fucking idiot*, he thought, *she just insulted you with a Lenin quote. Say something real for once.* Instead, he said, "I get it. I'm trying my best to get 'those with power' to do the right thing too. But no matter how bad I feel, the boat only turns so fast. And most of us aren't at the helm. I'm barely in the engine room."

"Then why should I let you waste my time?" Saga said, and she walked away, into the weaving crowd.

"Hey, wait!" Suddenly he was dashing after her, out of the pavilions and into a door-dense hallway. Without looking back Saga deftly dodged around a shuffling group in Ghanaian neotraditional dress and into a side event room. Noah tried to follow, but the prim ENGO bouncer spotted his pink badge and blocked him.

"At least I nailed the metaphor," Noah said.

First Friday

Noah fumed about the exchange with Saga all through Thursday—defending his honor in lengthy texts to Diya with perfect, too-late comebacks. It was, Diya pointed out, a good distraction from his actual problems: the apparent collapse of his career in diplomacy. But Friday brought Noah a big break. A pair of US delegates took ill with some globalized summer flu. As an experienced negotiator, however far out of favor, Noah was called in as an alternate.

He would be offering questions and, time permitting, giving a US reportback on the longitudinal health co-benefits of urban afforestation to the *Ad Hoc Working Group on the London Affirmations, Item 12: Matters Relating to Land Use Best Practices (LUBP) Referred to in Article 29*. ALA12, as it was called, was—after ALA4, of course—probably the most exciting track of negotiations taking place at this COP. "I'm not a land buff," he told Diya amid a string of excited car chase emojis, "but this is absolutely where the action is."

When he got the news, he speed-showered, put on the nicer of his two suits, powered through eggs with toast and maté at the hotel's buffet, dashed to the conference shuttle. He piled in next to a large Russian man sweating and sniffling. According to COP lore the first Friday of every conference was "incubation day." Germs caught on the plane or from other badge holders in the recycled air of Spaceship COP didn't blossom into colds and stomach bugs until the end of week one. Because they more often stayed both weeks and tended to get more strung out by the negotiations, delegates anecdotally got the worst of whatever went around each year. As the tired joke went: on the weekend, the activists go to the CAN Party, while the parties were busy on the can.

Normally Noah would be freaked out to be crammed in beside an obvious germ threat, but not today. Instead, he manically flipped between revisions while listening to an automated podcast of curated takes from pundits following ALA12. The informal consultation was not scheduled to be anything more than prepared interventions, but if he demonstrated expertise, it would be hard for Marta to shake him out of the rest of the ALA12 negotiations entirely—especially if the item got contentious.

And indeed, ALA12 proved very sticky. Different farming practices and urban development patterns would receive different levels of credit in the carbon accounting scheme being developed in the London rulebook. Those differences had economic implications for countries predisposed to certain crops who didn't want to retrofit their agricultural infrastructure. So, despite the thoroughly documented recommendations from the IPCC and affiliated researchers, the actual credit system would be determined by the same kind of bargaining that went into any trade deal.

Meanwhile, a few of the more recalcitrant parties— the holdout petrostates, of which America was a reluctantly reformed ex-member—were keen to make sure the ALA12 negotiations sapped attention and time away from items more directly hostile to fossil interests. So, the mild disagreement expressed in country position statements was enough to make the meeting sprawl into one of the closed "informal informal" sessions where much of the real debate and horse-trading at the COP happened. Noah happily went along with this. It felt good to stand his country name block vertical and be called upon as "United States." Most of what he said was suggested to him by remote teams who had computed the best negotiatory routes to get the text the US wanted. But like an old-timey live news anchor reading a teleprompter, Noah was drunk with power knowing that he could, if he decided to, say anything he wanted.

Then he got very lucky. A delegate from East Eswatini took him aside for a "triple-informal"—cubing, in COP jargon. East Eswatini was half the size of Delaware and twice as committed to global tax fraud. The delegate was young, clearly looking for bigger opportunities to come out of his participation in the negotiations. He needed a personal favor—help getting a totally legitimate shipment of electronic cigars from his brother out of customs lockup at LaGuardia—and he was willing to chat up the other members of the Least Developed Countries group about forming a like-minded issue coalition to break through the rice-versus-millet impasse brewing in ALA12.

"You know, I know just the person to talk to," Noah said, a friendly hand on the delegate's shoulder. Ten minutes later, when the session broke for coffee, he texted Diya, "Diplomacy is back, baby! Awoouu!"

Coming out of the negotiations for dinner, Noah glad-handed his way through the other pink badges, turning a few professional nods into appreciative smiles. Loss and damage issues aside, it was a decent year to be American at the COP. In the sixty-year history of the convention, the US had played the villain, the swaggering savior, the apathetic doomer, the hard-to-get lover, the tantrum-throwing child-god. This year the lapsed superpower was cultivating the air of a newly recommitted partner, returning to the fold bearing gifts, and Noah was happy to slip into this role.

As he moved toward the exit, he found himself in a swirling crowd, half waiting in line to get badged out of the venue by security, half just lingering, stroking their phones, swaying with the far-off focus of those on subvocal calls. Then he heard it: soaring John Williams horns and strings, with the badly sung lyrics every COP veteran knew. "Fossil of the day! Fossil of the day! Who is bad . . . who is worse?"

For decades the activist contingent at COP, led sometimes by the Climate Action Network and sometimes by rival ENGOs like GURR!, performed a mocking awards ceremony at the end of each day of negotiations, calling out whatever country they felt was holding up their vision of just and proper climate action. It was cheesy, but it was tradition.

He shuffled closer to get a better view. There was Saga, conducting the crowd in the familiar song, a mask over her eyes and a dancing cartoon skeleton projected onto her dark clothes. Behind her three grinning youths danced in clunky dinosaur costumes. Noah felt a twinge of frustration seeing Saga again, but he was in a generous mood. Of course he could appreciate her position, even if he disagreed. She just wasn't where he was, hadn't been in the informal-informal-informal, and so couldn't see how things really worked. No hard feelings, he thought.

Saga launched into the familiar monologue. "Each day we give this award to the country that is the *best* at being the *worst*, that does the *most* to do the *least*," she said. Noah knew the rap, but he'd never heard Saga recite it before. She was good, and her jokes were better than the usual CAN fare. The gathering crowd, big for a Friday, was getting into it.

She announced the third-place winner—Pakistan, which had given signs of backtracking on the ambition they had professed during the Pagniniig Colloquies. The teen stegosaurus handed the trophy to a Pakistani activist, who accepted on behalf of her country with sarcastic shame. Second place went to next-door Uruguay, whose president had apparently chided Argentina's climate failings without admitting his own—"people in glass greenhouses shouldn't throw stones." Noah vaguely remembered that Argentina had been top fossil on the first day of this COP, a cheeky move by GURR!.

"For our first-place fossil, you all know them, you love to hate them and hate to love them," Saga called, working up the crowd. This was the Valkyrie he'd caught a glimpse of, Noah thought, the natural performer. She did a drumroll on her thighs. "It's the United States! Give it up, everybody!"

Noah blanched. Not quite the endnote he wanted for his triumphant return to the negotiating table.

"Who here can accept this great dishonor on behalf of the good ole US of A?" Saga called. She scanned the crowd, and for a second Noah felt her eyes lock onto his. Then a gangly man Noah didn't recognize stepped out of the crowd and took the plastic gold trophy from the T-rex.

"America, we're here to make friends, y'all," the man said in a fake folksy accent. "So long as you don't hold us accountable for anything that happened in the past, we will happily lead the free world into a brighter, greener tomorrow!"

"But what about the refugees you've turned away?" Saga asked. "Some of the rafts you've droned had legitimate grounds to ask for climate asylum. Are you going to deal with the toxic culture in your Coast Guard?"

"A rising tide lifts all boats. And rising seas . . . well, that's even better!" the American said. "If you don't have a boat, don't worry! Pretty soon you'll be far enough underwater that we can forget all about you!"

Noah turned away, found the exit. America getting fossilized was nothing new, he thought. Happened every year. Sure, Saga might be gunning for the US a bit harder than normal. But she was mad about policy, she said so herself. Certainly no one could blame any of this on him.

First Saturday

Noah was wrong. The next day Marta Tolmbly finally arrived. After the morning huddle, she pulled Noah aside.

"Campbell, I'll be straight with you," she said. "There's going to be some bus throwing."

"You're throwing me under the bus?" Noah said, crestfallen.

"No, I'm fucking throwing a bus at you like I'm the goddamn Hulk. That's my level of anger about this."

Marta was tiny, but somehow seemed to move with incredible muscularity under her atemporally in-style pantsuits. She'd come to the State Department from the military, and sometimes Noah thought he saw the hitch of a robotic prosthesis in her step—but of course it was impossible to tell anymore.

"I take full responsibility for the ECO article," Noah said. "But the fossil thing, they were more mad at those Coast Guard fanatics than at anything I said!"

"I'd keep that sentiment to yourself, Campbell," Marta said. "Our commander-in-chief had an illustrious career in the Coast Guard. He is not keen to have his reform efforts critiqued on the international stage."

"Well, no, that would be unseemly," Noah admitted.

"You may recall we had a midterm a few weeks ago," Marta continued. "It went pretty badly for the president. So, our new objective is to leave Buenos Aires with a narrative that gets Americans excited about the president's Clean New Deal proposal. What we absolutely don't want is a series of PR debacles that gives the president's opponents an excuse to cast the London Affirmations—and by extension the whole cleanup economy agenda—as just . . . what do they call it? 'Weather welfare for foreigners.'"

"Ma'am, I am totally on board for that," Noah agreed. "Quiet, classy diplomacy, no media spectacle. That's exactly what I'm all about! You should've seen me yesterday in ALA12—"

"What are you going to do about GURR!?" his boss interrupted. "Aside from hounding us, they freaked out

Argentina enough that the intelligence secretary asked us to crack their network. That's not the kind of request I like to say either no or yes to."

"Great question. I've already devised several strategies I think you'll enjoy." Noah had done no such thing, but he believed that talking was a form of thinking. The bullshit that was about to come out of his mouth was likely just as good as whatever he might have prepared beforehand.

"This Saga Lindgren person," Marta said. "What's your strategy for her?"

"We got off on the wrong foot. It happens. She's leaving tomorrow. I'll stay out of her way, and things should blow over."

"You'll do no such thing," Marta said. "You need to find her and apologize to her."

"Today?" Noah had hoped to maneuver this conversation back toward his participation in the ALA12 negotiations.

"Yes, today," Marta snapped. "Look, you and I both know there's only so much money we can spend on climate change. And if we get sucked into writing a check every time something bad happens, we'll never have the funds to do the big work of fixing the world for good. You and I are paid to make that calculation, because the United States of America has public education and daycare subsidies and all that jazz. Our bean counters need us to survive long enough to see a return on that intergenerational investment. Activists like Lindgren file yearly impact reports, so there's no incentive to make the hard choices for a payoff decades away."

"All due respect," Noah said. "That might be oversimplifying her decision matrix."

Marta rubbed the bridge of her nose. "I just got off the red-eye. I've got some nonsense emergency meeting with the secretary general in ten minutes. You want to remain on this delegation? Convince Lindgren that GURR! needs to choose between repair and reparations this COP—if they want a shot at getting either."

So Noah spent the morning scouring the COP venue for Saga. This was easier said than done, with tens of thousands of people churning in and out of dozens of negotiation sessions, plenaries, side events, workshops, press conferences, meetups, gallery exhibitions, startup launches, film premieres, academic forums, panel discussions, activist actions, musical performances, teach-ins, and sustainable cooking classes. It was easy to bump into acquaintances in passing, but hard to actually find someone deliberately if you weren't in direct communication—and Saga wasn't returning his messages.

So while he tried to triangulate what sessions Saga might be interested in, he obsessively watched GURR!'s social feeds, blocking other notifications so he could focus. Around noon a loop of video popped up showing a cluster of activists posing with GURR! signs in front of a white statue of nude women, bursting forth from a watery veil. "Meet us at Aurora to demand repair and reparations!" the caption said. There was no GPS tag, and their faces were blacked out, their silhouette lines glitched—an op-sec thing, he guessed. But he was sure Saga was among them.

He dashed out of the COP, ignoring the mumbles of the guard who badged him out, past the shuttle pickup spot where polite but firm UN staffers were hustling delegates onto a herd of buses. Some kind of party sightseeing expedition, from one of the dozens of emails he'd batch-deleted? He tapped furiously on his phone to hail a car, swiping away four autonomous cube-cabs before he found a taxi with a human driver. It rolled up four minutes later, a yellow four-door, ribbed with custom LED lighting.

"You gotta enter a destination, chabón," the driver said when Noah climbed in.

"I want to go here," Noah said. He held up the video, zoomed in on the statue. "Do you know where that is?"

"You want a tour, I can do that for you, but you gotta pay on the other app for that. Also I'm going off the clock soon. Gotta get inside, you know?"

Noah glanced at the ID laminated on the dashboard. "Look, Luis, yeah? How about you get off the clock right now, and I'll pay you direct. Skip other pickups and get me to this statue pronto."

He pulled up a generous sum on his phone. Luis the driver sized him up, craned his neck to look out the windshield at the sunny sky, then reached back to tap his phone to Noah's.

"You got it, chabón," Luis said. "But you picked a bad time to try to get anywhere fast. They're closing down streets for the march."

Noah groaned. Of course. The climate march was today, as it was every first Saturday of every COP. He'd never marched himself, but he'd seen it from afar the couple of years local police had allowed the protestors to get close to the venue.

"I need to find a woman at the march," he said. "Just get me as close as you can, as fast as you can."

Luis nodded, and the car ripped away from the curb. "If this is a love thing, you might want to wait, my friend. Gonna rain soon, and anyway la policía aren't taking no shit today. Let your woman do the march. It will calm her down. Next week bring flowers. Does the trick every time."

Luis had a wide face and long hair in a messy, feminine topknot, his build slight beneath a baggy e-sports jersey. He looked young, maybe midtwenties. Noah remembered his own impeccable romantic wisdom at that age.

"No, it's just a work thing," Noah said. "And she'll be gone next week."

"Way you say that, chabón, doesn't sound like just a work thing."

"Just drive, kid," Noah sighed.

He stared out the window, his first time really seeing Buenos Aires outside the COP. Buildings were mostly squat, concrete boxes, their tan paint storm-stained but mostly free of advertising. Pedestrians hustled by shoulder-to-shoulder. Old women retrieved their laundry from caged balconies. Traffic was orderly but cramped. Ahead Noah saw a blockade of vans, black and electric blue, topped with lazy-spinning siren lights. Luis turned into an alley, which blossomed with graffiti murals.

It was sticky humid. Noah pulled his suit jacket off, cuffed his sleeves. He couldn't remember the last time

he'd attended a real march, if ever. His parents had pic-
tures of taking him to protests as a baby, but the culture
war was too hot to be safe by the time he was a teenager.
Now he felt painfully aware of just how much he looked
the part of a thirty-seven-year-old gringo technocrat.

Luis wound them through side streets and cut
through parking lots, routing around the urban damage of the concentrated police presence. Sometimes he
cursed, zoomed and tilted with deft fingers on the cab's
glitchy map screen, flicked away pop-up warnings Noah
couldn't read. Eventually they pulled to a halt in a tagged
and molding alleyway—once a wide street, narrowed
by densification. Luis pointed at a passage between two
apartment stacks.

"I think this is as close as we can get, chabón," Luis
said, "unless you want to wait through a checkpoint.
Down that way, past the pool, the stone building has a
side door that's usually open. Go out the front, and you'll
be across from the Parque Centenario. Your statue is *La
Aurora*, on the lake's south side. Take this."

The young man pulled a cheaply printed umbrella out
of the glovebox, handed it to Noah.

"Storm blowing in, my friend. Don't know how long
this march will go, you know? Rapido, chabón."

"Sure, rapido," Noah agreed. He took hold of the umbrella, but Luis didn't let go.

"Buenos Aires is not for beginners, my friend," Luis
said, serious now. Noah looked him in the eye and nodded. Luis released the umbrella.

Self-conscious that he was semi-trespassing, Noah
jogged up a few steps and between the buildings into a

raised, informal courtyard. At first he didn't see a pool, and panicked, but then he saw the garden: perfectly rectangular, edged with tile, overflowing with carrot greens, lettuce, squash vines. Luis's car must have been using an old satellite model, from before this neighborhood had downshifted during one of the last several financial crises. Clotheslines were strung between the buildings. A filter-fitted rain barrel stood under a drainpipe. Along the walls of the courtyard were boysenberry trellises, and in one corner a wire chicken coop clucked and cooed. Above him tenants were pulling in flimsy photovoltaic panels that had been stuck out of windows to catch the noon sun.

Noah edged his way through the tangles of permacultural apparatus, to the gray brick building with a red back door. No keypad; he tried the knob, and slipped inside. The hallway beyond was dim, and when he pushed through the next door, the sky-blue open space seemed to assault him. He was in a cathedral sanctuary, stained-glassed and marble-columned. A pale, unhappy Jesus regarded him from the cross.

Noah walked past empty rows of wooden pews toward the exit. Almost empty—he caught the eye of a knot of youths, whispering, protest signs in their laps. He pushed out into the electric air.

The circular park across the street was mobbed by angry color. Thousands milled around, rocking with anticipation. From the church steps, Noah could see clusters of bobbing signs trickling out to the east. The march was already starting. He hurried and merged into the crowd.

The climate march was as traditional as the opening ceremony, but Noah had never seen it up close before. Now that he was inside—hearing the half-rhythmic protest chants, seeing the banners and T-shirt blocs, the loud costumes and puppets, the writhing columns of projection fog, smelling the sweat and weed and glue and fear—he realized what a different experience of the COP these activists and locals must have each year.

And yet, text dominated here as well. Everywhere were signs and shirts with slogans, manifestos, demands, org names, jokes, memes held grinning on cardboard. Dancing between these were arcane symbols: the neutral circles of the climate sign, the broken hourglass of Extinction Reckoning, many boiling cartoon thermometers, the tornado emoji, the burning earth, the lightning-struck tower of the tarot.

Half-dancing, half-shoving, Noah made his way to the water, followed it until he found *La Aurora*. In person he could see that the veil of the white goddess concealed a carved-stone workingman and two plow oxen. But Saga and the GURR! contingent were nowhere to be seen.

Noah thought about heading back to the church, back to the COP or his hotel, getting out of whatever weather was bearing down, giving up on this wild goose chase and throwing himself on Marta's mercy. Instead, he plunged ahead into the march, trying to move with and faster than the human current. He began to pick out amoeboid blocs, singing or shouting as one. They shared sign-themes and sometimes shirt colors, were kept together and steady-paced by dancing protest marshals. Some carried long, inflatable tubes, flashing with video projected from within. Others held tall tapestries that billowed like sails in the swift South Atlantic breeze, tugging against the muscles of straining youth. Noah dodged and weaved around all these, searching for signs of GURR!'s angry polar bear mascot or, failing that, Saga's shimmer of yellow hair.

Around him some of the marchers were tucking their signs between their legs to pull rain gear out of backpacks. There were indeed marble-textured clouds now filling the sky overhead, moving with uncanny speed. He heard snippets of urgent conversations. "Should we go?" "No way, they make up something like this every year." "They say it's turning hard." "We've probably got hours, though. You want to go back through that humiliating checkpoint so soon? Leave if you want. Cops don't look worried."

Noah had no way of telling if this last was true. Most of the police lining the street on either side of the march had helmets with black visors that covered their faces. A few had pulled-down balaclavas. All wore bulky riot gear, clubs and zipties dangling from their belts. Some carried assault rifles. Others held weapons Noah didn't recognize that looked better suited to pest control. As he watched, one cop pulled out a sheet of anime stickers and pasted them over his body cam and badge number.

This show of force was new and unsettling for Noah. Why the stormtrooper intimidation tactics? For that matter, why bring in the enormous, tank-like vehicles that were parked at every turn of the march, topped with ominous black disks? In America he knew the police were Problematic, but he also had no immediate,

personal fear of them. At the COP he was used to treat-
ing the police and rent-a-cop security the same way he
did blue-badged UN staff. He assumed that his status
as a delegate, an American, a government official would
earn him some institutional deference—but, a nervous
voice in the back of his mind whispered, these individual
policía didn't look so discerning.

Noah pushed forward faster, eager to at least get to the
front of the march. Then he could leave, tell Marta—and
himself—that he'd done his best. Maybe Saga wasn't even
here. She was probably back at the COP, or at her hostel,
packing before the CAN party, checking her things for
stowaway bedbugs.

But she was here. He saw her then, yellow hair dancing near the very head of the human snake. Gone were
her spartan business clothes; instead she wore a tattered,
dramatic protest costume, blue and green and black, and
shook a pole topped by a comically large papier-mâché
polar bear paw. He rushed to her, flooding with relief
and new anxiety. He didn't actually know what he was
going to say.

"Saga, hi!" he said, bright and casual as he could. "Fan-
cy, uh, meeting you here."

She turned and stared at him. Her face was painted
an aquatic blue, topped by waves that rose to the level
of her eyes, submerging her.

"No costume?" Saga said. "No sign?"

"Yeah, well, gotta keep my hands free for this umbrella," Noah said, holding up Luis's gift. "Just here to, you
know, put my body in the gears of the evil machine!"

This, he knew, wasn't fooling her, or the other GURR! marchers looking at him with skepticism. He swallowed.

"Look, I wanted to talk to you," he said. "Apologize. For hitting on you while you were trying to do politics. Or trying to get juice, I guess. I'm sorry."

Around them the protest thumped and wailed. The murmur of wind grew louder.

"Why," Saga said, "do you keep following me?"

He wanted to say: because he felt attracted to her, connected to her, challenged by her, seen by her more in their few brief encounters than by anyone else he'd met in a long time. It was a completely irrational feeling, he knew, erupting into his psyche in a cloud of privilege, loneliness, career frustration, and climate despair. He was sure there was something stunted in him, a twisted organ, corrupted by power and patriarchy and everything else that asked him to look away from the intensity of others' needs. But, in that same thought, he was also just as sure that the stunted thing was in everyone, each of them looking away from something else.

He couldn't say any of that to her, not here, in the eye of whatever storm was brewing around them. So instead he said, "I'm being pushed out. Domestic politics. I'm the only one who doesn't toe the company line on financing loss and damage, because I've been around long enough, know enough people, that I can call bullshit when I see it. Internally, that is, but they don't like that. The president is running from his enemies, scared of the media, trying to streak across the quad without anyone noticing the emperor has no clothes."

He winced at his own metaphor, but Saga didn't raise her wave-lapped eyebrows. He soldiered on.

"They don't want to hear any talk at this COP about loss and damage, reparations, anything like that. Which means," he paused, feeling like a weasel, small and cunning and proud, "you have them by the balls."

Saga smiled then. "So, you finally want to negotiate?" she said. "Well, let's negotiate. That is what you came to Buenos Aires to do, yes?"

Wind gusted, snatched a word from his mouth. He wondered what it would have been. When the breeze died down, he tried again.

"I can help you. I can convince them that GURR! has a whole media shame campaign locked and loaded, ready to target exactly the people State is scared of with exactly the messaging they fear. They'll bend over backwards ramping up ambition on sequestration to keep that contained, way past what they'd like to promise. And if I get the chance, I can work triple-informals to box them into long-term commitments on reparations. Not this COP, but I can set things in motion. I'm good at this shit. But you have to back off next week and let me work my magic. I have to keep my job to be of any use to you."

He could see her calculating, trying to decide if he could deliver what he promised, and how much she would lose if he couldn't.

"How, precisely, would you be of use to me?" Saga asked.

"With me on your side, you'll know just what pundits to hit to get the State Department's ear, just how to work the news algorithms. We'll play them, keep them off balance, this year and beyond. I'll tell you when GURR! should rattle the cage, and when to dangle an olive branch. They'll think you're preternaturally scary, and they'll think I'm the only one who can keep you in check. The harder they squeeze us both, the more incentive we have to work together."

Another too-long moment of her sizing him up. "Maybe we can help each other," she said carefully. "But, listen, I'm not going to date you. I can't. Not until the world is less . . ." She waved around them. "Like this."

"Nothing personal," Noah agreed. It was bittersweet, to get this glimpse of hope that she might just share his attraction, wrapped in an ultimatum that it could never work. He stuck out his hand. "Just politics."

She laughed her laugh then, at his seriousness, and lowered her pole-mounted polar bear paw to meet his palm. They shook.

Noah opened his mouth to say something else—he had no idea what—but again his words were washed away by the wind, and then by something else. A colossal force seemed to batter him. It took him a second to realize it was noise. He covered his ears. Saga and everyone around him did the same. Then the attack formed into words, piercing through his hands and into his skull.

"Dispersen!"

It was the police, using some sonic weapon. Noah had never heard one before. He felt nauseous.

"Dispersen!"

The cops around them had riot shields up, clubs in hand. They stepped forward in awful unison. Noah felt buzzing on his leg and automatically pulled out his phone. Priority notifications from the city, the police, the COP—all saying, with varying degrees of implied violence: leave the march, get inside now. Below those was a string of blast messages from Marta, advising the whole delegation on which shelters to head for, and one loud text from Diya—"HOLY SHIT! HELENE IS COMING UP THE RIO!"

Noah did not intend to argue with any of those messages. Around him, however, the story was more complicated. Some protesters had lowered their banners, stood on tiptoes looking for a route out, but others had turned their attention to the cops, were shaking signs and fists, shouting "Fuck you, pigs!" Rhythm boiled amidst more blasts of the sonic weapon: "Whose streets?" "Dispersen!" "Our streets!" "Dispersen!"

But they couldn't disperse, Noah realized, because the cops weren't letting them. Police had enclosed the head of the snake and lined the sidewalks shield-to-shield. Anyone who got too close was shoved back. In his noise-rattled brain, Noah couldn't parse whether this was a logistical error or some deliberate, doublespeaking provocation. The crowd around them began to compress as the marchers further back pushed ahead, unaware that forward wasn't the way out.

"We have to go back through the crowd while there's still room," Noah said to Saga, half-yelling.

Saga ignored him, turning instead to her GURR! comrades. "We must break their line! Make a hole for people to escape! Tara, Shi Ann, spread out, find ER, anyone with martyr tats. Get those tubers ready to point the way."

She had made the same calculation about the mounting pressure of bodies as he had, but had come to a different conclusion. Noah watched as she deployed her troops, working her phone, reaching out, he assumed, to allies elsewhere in the march.

"Hold this steady," she said, and she pressed her phone into his hands as its projector flicked on, mapping a dim, drone's-eye-view of the area onto the dirty street. The software looked like a mix between a military strategy game and a project management system. He tried not to move, and Saga and the remaining GURR! activists scuffed at the interface with their shoes, assigning tasks and positions to allies as they popped up as colored dots. It all took less than two minutes, but this time felt endless in the tumult of police commands and the growing cacophony of the crowd.

Then Saga snatched her phone back and held her pole like a spear.

"Let's go," she said, and she began pushing her way toward the cops, now banging on their shields. Noah stumbled after and sensed, as though through some mass proprioception, the crowd shift and reorient with her. Around him dozens were moving in the same direction—not everyone, but enough that more heads turned, bodies faced toward this new inertia. The long, inflated videotubes and the ghosts of projection smoke began to switch, one by one, into coordinated arrows, "this way" signs, and gifs of stampeding buffalo.

It was then that Noah remembered Marta's comment that morning. "Wait," he said and lurched to catch up with Saga. "Wait!"

But when his hand found her shoulder, it was already too late. The police knew where they were going and had converged to meet them. Instead of a thin line to break, they met a mass of riot cops, crashing in toward them. Just yards ahead Noah saw plastic shields smack against cardboard signs. A truncheon came down on the neck of a thin, gray-haired woman. A man in some kind of indigenous garb stepped forward to grab her and caught a boot to the face. Enraged, the crowd slammed back at the cops, and there were tussles for shields, shoulders put into the breach.

There were thousands of policía and thousands more protesters. Noah had no idea who would win, but he knew he didn't want to be in the middle of it. In front of him Saga turned, her phone back out, sending commands to redirect the movement of the crowd.

"It's not going to work!" he said, loud in her ear. "They're in your network!"

Saga looked up at him, eyes startled. There was fear there, for the first time. Rain began to fleck down upon her face.

"Come on," he pleaded. "You can't help them. This is Sisyphus! News algos won't top police beatdowns with a hurricane coming. We have to go back. No more politics is going to happen today."

She looked around, then up at the darkened sky, then nodded. She took Noah by the hand. Even in the chaos, it felt electric to Noah. Together they moved against the heaving crowd.

They ran, Saga gathering up her GURR! comrades with shouts and glances, Noah trying to recall his route through the march just minutes earlier, steering them around still-held banners and huddled blocs. For a few moments it was easy, the wind whipping at their backs, dodging and weaving, splitting with Saga and coming back together—a strange sport. Then the rain began in earnest.

The first moments were almost a relief—Noah was so sticky with sweat and fear. But then he was soaked heavy and half-blind, his feet slipping. He tried to find the latch to open Luis's umbrella. Around him what cohesion remained in the march disintegrated. Inflated videotubes sagged to the ground. Projections turned nightmarish in rain-pocked smoke. Individuals dropped their signs, and blocs began to scatter.

By some violent logic Noah didn't understand, the rain seemed to unchain the police as well. Cops advanced into the crowd, no longer a formation, now moving like hunters, raised clubs seeking disobedient flesh. He shrunk back from a masked demon, moving on him with a long-barreled spray gun, spewing tear gas impotently into the rain-washed air. Saga was gone—he didn't know how—and another cop appeared at his flank, shield swinging at his legs, aiming to trip him. Noah stumbled, tried to grab and raise his UN badge, but the cops didn't care. He fell to his hands and knees, and a club smacked into his ribs.

Noah yelped in pain and choked as the gas nozzle got closer. Somehow the umbrella was still in his hand; he swung it wildly toward his attacker. It connected and popped open. The cop beating him grunted and sprawled on the ground. Then, ahead of him, he saw Saga's white pole-paw, reaching toward him. He scrambled and grabbed it, and she hauled him up.

They ran again, alone together now, back through the turns of the march, keeping to the middle of the flailing crowd, looking for a way to clear the street. What had taken an hour to march now sprinted by in minutes. Ahead Noah sensed the wide opening of the Parque Centenario, now stirring with the dregs of the protest.

"That's an Extinction Reckoning cadre," Saga said, pointing at an orange-clad group in tight formation holding back a few cops with broken-hourglass flags. "They can help us run a checkpoint."

"No!" Noah said, marveling at her daring, thinking again of the Valkyrie. "No, this way."

He pulled her left, hoping he could find his way through the slapping sheets of horizontal rain. The wind was howling now, though Noah knew the storm was just beginning. He looked left and right, and saw it: an open gate, a gray building, a red door. Noah slammed against the wood, found the handle, slammed again. They fell into the church.

Second Sunday

Noah and Saga spent the night huddled in the back offices, eating communion wafers and sipping wine without getting drunk. Mostly they talked, a loose, empty banter that promised little and meant nothing, but still had that frisson of intimacy. Once Noah had felt an almost overwhelming urge to leap across the small room and kiss her, but he'd stayed put. He felt good about that.

Eventually some animal instinct to sleep through a storm kicked in, and they passed out on the spare pews.

When morning came, they crawled out of the back rooms. The cathedral sanctuary was strewn with shards of broken stained glass, pooled by ovals of damp where the rain blew in. They tried the front entrance, but a pile of debris and a torn awning had wedged into the doorframe. Through the crack, Noah could see the park, littered with downed streetlights and abandoned protest paraphernalia—dirtied and swirled by the shallow flood.

Saga went to kick at the door, but Noah stopped her, led her through the church to the back entrance, and out into the downshifted courtyard he'd first snuck in through. Here half the trellises had fallen over, a few of the clotheslines had snapped, but otherwise not much was changed. They drank thirstily from the overfull rain barrel.

"This place seems to have come out all right," Noah said. "Look, the chickens survived."

"This will be the exception," Saga said. "Trust me. I've seen it. Cities unprepared, hit by storms that were never supposed to happen. Except they happen every year now."

"When's your flight?" Noah asked. "You were eager to leave. Are you going to stay?"

"I'll probably stay. Flights will be canceled anyway."

Noah resisted the urge to ask if she was free for dinner. Instead, he said, "Do you think this will change anything? The COP getting hit by a neverstorm?"

"No. Do you?"

"No. Well, maybe. Certainly gives your lot some good, metaphorical ammo."

"Don't you mean 'our lot'?" Saga said. "We have a deal."

"You never stop, do you?"

Saga gave him another of her canny appraisals. This time, she looked tired. "Noah, if you could actually make something happen at the COP, what would you want?"

It was a question Noah hadn't much considered for a few years. He thought about it while he strolled around the garden, kneeling to check on the carrots. He went to the wall and righted the downed trellises. He felt a sudden, deep gratitude for the courtyard. It wasn't even his to claim, but nonetheless it was the one spot he'd felt at peace this entire week.

"I don't know," he said. "I like places like this. My family built a lot of them in California. Pockets of resilience, where people can ride out the storms. I guess I always wanted to work on something bigger than setting up rain barrels. But more pockets like this seem like something we'll probably need."

"And the people who don't have a pocket?"

"Well, I suppose we better do something for them too," Noah said.

He beckoned, pointed toward the exit. They walked out of the courtyard, and into the damage.

Shared Socioeconomic Pathway 5
Taking the Highway—Fossil-Fueled Development

TOO FAST TO FAIL

First Sunday

To pick up his COP badge, Luis wore his blue anti-pleated H-Zoo suit with the graphic paisley lining that lit up neon in product lighting. This he paired with studded vegan high-tops from the Jimmy Choo–sponsored dancewear imprint Imm0biles, which he borrowed from his artsy younger sister. He topped the fit off with an exquisite bun pattern licensed from a Canadian stylist for whom he'd done some brief but lucrative modeling work.

"Isn't it gauche to wear a Taiwanese jacket with Malaysian shoes to a diplomatic event?" Marcela asked as he pulled on her shoes. "Given the unfortunate brand tensions between those two countries, I mean."

"Naw, sis, I've done my homework," Luis said. "New boss was a pre-alpha influencer for Imm0biles' parent platform, Travesty Dot Club. If I want to come out of this gig as more than a booth boy, I gotta demonstrate some serious brand awareness!"

So fitted, Luis gassed up his freshly leased Mazda and gunned down 25 de Mayo to 9 de Julio, then up toward the convention district in Recoleta. It was a beautiful day! The sun was shining hot, so he cranked his car's AC, that life-giving breath. Along the highway he watched the projections on the spinning air purifiers dance with the logos of the road's new December sponsors—Fiat, Alibaba, Flovent, Haliburton, ChevrExxon, and, of course, the UNFCCC. He parked at the Four Seasons, splurging on a valet.

The official event didn't start until tomorrow, but the expo center buzzed with crews doing setup, camera drones shooting B-roll for product-story convention logs, venture scouts munching cheese and prosciutto at pitch-preview socials. Luis, strutting with excitement, threaded around the workers and picked up his silver-banded vendor badge, then passed through security and into the main exhibit hall.

Thousands of booths and pavilions gridded the cavernous space. There were displays from every major company in the world, next to booths for hedge funds and public prosperity funds; hundreds of startups bumping shoulders with half as many micro-angels. It was an incestuous mixing of enterprise and capital, all looking to either find or *be* the next product or platform to break new market ground in one of the three R's named in the UN's Growth for Earth Goals: risk, recapture, redevelopment.

On the ceiling above, a massive projection showed a sit-down between the new incoming COP president—the CEO of the Central Bank of Argentina—and an elegant foreign businesswoman with gray-streaked hair. As he walked, Luis tuned into the audio on the COP Plenary Podcast livestream.

"Our motto this COP is 'Dynamic Planet, Dynamic Markets,'" the COP president was saying. "Can you tell us what that means for your investment ambition, Ms. Kapoor?"

"To me, it means we need to spend aggressively to keep up with the weather," the woman replied. "There's no use letting good capital sit and rot, so I'm going to be out on the floor tomorrow with cash in hand, looking to take the most aggressive solutions from zero to hero!"

"Wonderful! That's what we like to hear!"

Luis licked his lips. For two weeks the booming climate sector was pouring a small sea of money and opportunity into Buenos Aires. Luis was stoked. He was ready to dip his toe into that water. In fact, he wanted to strip down to his Armani Exchange thong and dive straight in.

He zigzagged to the coordinates his boss had given him, smiling at the foreigners and giving dap to the few locals he recognized from the BA green biz scene. The booth was smaller than he'd hoped, but shapely, probably fabbed and set up by some high-end convention services firm. The logo was a trendy swoop of shipbuilder schematics—but he'd have to rethink his outfits this week given the color scheme. Those blues were *definitely* darker than the ones in his onboarding packet, at least in this lighting.

Next to the logo was the company's name and slogan. *Ark: Build It Before the Rain.*

First Monday

Luis met his new boss the next morning. Noah Campbell was a beach-blond American entrepreneur with the packed-on physique of someone who overdid the protein powder and underdid the actual exercise.

"Wow, you really blend, my dude," Noah said, surveying Luis's cobalt-blue fit. "You and the booth look like a package set! Nice shoes, though. Are those Swiss?"

"They're Imm0biles," Luis said, unsure how much further to name drop. "Limited edition!"

"Never heard of them," Noah said.

Luis took this in stride, pivoted back to expanding his role with Ark. Noah, Ark's Founding Chief Visionary Officer and sole full-time employee, had hired him via a local staffing firm to man the booth and hand out Ark marketing literature. When the accounts rep contacted him, Luis had bribed her to look up who else Noah had contracted. No one. Which meant Noah was either still looking for talent or was foolishly trying to go this COP alone and would soon find himself in need of help.

"I think it's important to dress in harmony with our brand identity," Luis said. "Especially if you want me to get out of the booth to drum up contacts. Potential customers need to be able to recognize me as an Ark affiliate from a distance!"

"Investors, not customers," Noah corrected. "My dude, we are not selling a product. We're inviting peo-

ple to build their future with us—to be part of our sur-
thrive-all family! Every customer gets a stake in Ark,
and every investor gets access to our transformative
future management tools. That way *our* success is *their*
success."

"Boss, that's an incredible pitch," Luis said, heart-
feltedly. "I'd love to have a conversation about being a
real partner for you in making this vision a reality!"

"Okay, let me ask you this, my dude," Noah said. "Are
you willing to be cougar bait to score a big sale? You
willing to take the pounce?"

"What?" Luis said. "Sorry, my English . . ."

"Older, rich ladies, dude," Noah said. "The heavi-
est-investing demographic four COPs running has
been top-bracket females ages forty-five to sixty-five.
So what I'm asking is—you know what. I better stop.
I don't know what workplace harassment statutes are
like in Argentina. All I'm saying is, just think about that
statistic. Soak it in."

First Wednesday

Luis got his chance a couple of days later, working the
afternoon crowd from his booth. He was handing out
Ark pamphlets to strolling suits when a video crew de-
scended upon him.

"Wow! Blue on blue on blue! You're like a blueberry
in a blue tart!"

It was the woman from that Sunday ceiling interview,
Diya Kapoor. She leaned into the Ark booth, snaking her
arm around Luis's ribs and turning so they both faced

the camera wand she held at arm's length. She smiled winningly.

"Tell the viewers on stream where you're from, Blue," Diya said, bringing the wand in a little closer.

She looked different in person: no longer filtered by makeup algorithms, no longer twenty meters tall. She wore a blood-red sari—a Hugo Boss number, if Luis wasn't mistaken—and strolled through the expo backed by a colorful retinue of flunkies and retainers. One of those held up an extra-long tablet, on which scrolled the lively chat commentary of Diya's followers.

At his boss's insistence, Luis had reviewed the profiles and investment habits of a couple dozen—to use Noah's term—"COP Cougars." Diya was perhaps the most prolific and generous of these high-level angel investors, her life story fodder for half a dozen Bollywood makeover dramas. Once the nerdy black sheep of her star-studded filmi family, she had dropped out of university and turned into a major climate capital thinkfluencer, parlaying her family's fortune and media connections into a viral video persona the culture critics called "Anthony Bourdain meets Shark Tank meets Make-It-Rain Tremaine."

"Ms. Kapoor, I am from right here in Buenos Aires, born and raised." Luis flashed a vintage gang sign to the camera and switched to local Lunfardo slang. "Shout out to all my porteños hustling and grinding at COP60! Hasta las manos, but che che we made it!"

"A local boy! I love it!" Diya enthused. "Tell us about Ark. Is it an Argentinian firm?"

"Ark is totally global, which means we're also totally local," Luis said, recalling Noah's coaching. "We're about helping firms, individuals—everyone!—make smart choices about how to build their home-sweet-home *ready* for tomorrow's next-level superweather. Whether you are betting on water prices or fretting about sea-level surges, Ark uses state-of-the-art deep unlearning to predict and *manage* an ever-accelerating future! Ms. Kapoor, I promise this is an adaptation platform you'll want to invest in *and* use yourself."

"Wow! That's the best opening pitch I've heard all day," Diya said, half to him, half to her camera. "But a great pitch and a great product won't get anywhere without a great angle. My viewers and chat want a brand culture that gets them excited, a team they want to see win! What's your arc, Ark? Are you a scrappy underdog looking for your first big break? A rookie firm gunning for the crown? A visionary disruptor determined to change the game? As we say on my show every week, 'Don't have an angle, won't find an angel!'"

This had not been covered in the onboarding literature, and Noah seemed to change his story every time he breezed by with an updated sales rap. So, Luis decided to improvise.

"We're prophets with an actionable message about tomorrow. Climate adaptation is the biggest business in the world, but every year we see more major, low-likelihood weather events outside of traditional almanac patterns—so-called 'neverstorms.' When neverstorms hit, cities too often aren't prepared and haven't made the proper investments to capitalize on the market opportunities climate shakeups bring. Our story is those missed opportunities. We see amazing markets like Manila and Karachi not getting the redevelopment they deserve, and we think, 'We want to do it better!'"

Luis felt in the zone. Despite his discomfort with the sudden flock of cameras, he knew this was the big shot he came to the COP looking for.

"Blue, that gives me chills," Diya said. She took him by the chin and waggled his face at the camera. "The chat is clamoring to have you advance to the next round! Tell your founder I want to see *you* Friday night for the full deck pitch."

"I'll be there!" Luis said. He wondered how the older woman could parse the fast-scrolling blur of comments and reacts on her flunky's tablet.

"Wonderful! That's a wrap from day three of our weeklong COP60 special! Blue, can you help me with the sign-off? Hashtag BreakItToMakeIt. Hashtag MoneyToBlow. Don't forget to like, superlike, share, reshare and subscribe. And what else should they do if they want to see your next appearance, Blue?"

Luis tried not to roll his eyes at Diya's old-timey vlogger style, that over-the-top Gen Zennial enthusiasm born of the poorly managed social media addiction many older people still suffered from. He knew from dealing with his parents that the best thing to do was play along.

"Banners up!" he said, holding aloft his phone. "Give those noties full priority!"

"Love it!" Diya said. "We'll see you on Friday!"

And with that she whisked away, still chattering into her camera wand's ominous black bulb.

First Friday

"My dude, you look like a billion hypothetical bucks," Noah said, chewing his cheekmeat and admiring Luis's hourly rented vat-silk Berluti suit. "Not rich exactly, but, you know, future rich."

Noah and Luis waited in the tiny greenroom of the auditorium where Diya and her entourage were shooting the finals of her investment competition in front of a live studio audience. The venue was the MARQ Museum of Architecture and Design, next door to the main COP expo hall. Swelling wooden spheres hung from the ceiling, connected by rope bridges—an exhibit on trends in high altitude real estate. Apparently booking the place had involved a feisty bidding war with other financitainment players: TeamLiquid, CNBC, Rich Ass Mutants Dot Furry, the Trump-Kardashian-West Organization.

To Noah's mild jealousy, Diya had insisted that Luis be the one to give the pitch. So, they had abandoned their COP booth and spent the previous two days prepping Luis for the meeting, perfecting their slides and animations, running two-minute Monkey Ninja Bootcamp drills in case Diya demanded Luis complete a physical challenge.

All this was exactly how Luis had drawn up his dreams of "seizing COPportunity," as the flyers at the Future Business Leaders of Buenos Aires youth club meetings had said. Steering Ark into funding from Diya Kapoor would make him more than a booth boy—he'd be a real player in the climate tech scene. No more pouring maté for contact cards at pre-breakfast networking raves, no more borrowing clothes from his sister or begging for layaway at every boutique in town. Noah Campbell would owe him one, see his value add, maybe even make him a full partner. He'd pay off his debts, get his own place, and make some money before the next climate crash.

But now that it was in front of him, Luis felt a giddy apprehension.

"Boss," Luis said, nervousness creeping into his voice. "My uncle's Forbes life coach told him that the Kapoor family keeps a tight grip on the reins of the companies they portfolio. Are you sure you want to be getting in bed with them?"

"God, I hope she keeps a tight grip in bed," Noah said, picking at the now cold plate of lamb satay the production assistant had brought. "Diya Kapoor can put a full saddle and harness on me, as far as I'm concerned."

Noah had been like this since Luis had given him the good news: feeling confident, getting nervous, channeling that nervousness into sexual frustration, then using his libido to rally his confidence. Luis had seen this pattern before, back when he taught tabata classes for a summer, before getting the Peabody Energy scholarship that allowed him to attend business school. It was a temperament common to the founder class, which Luis thought left room for real advancement to those with the emotional intelligence to steer the flaring energies of an entrepreneur like Noah. But suddenly Luis wasn't just steering—he was at the head of the charge, being rushed into Diya's big cat maw.

"I'm serious!" Luis said. "You saw those desi toughs she's got down there. You better be able to deliver what I'm pitching!"

"Hey! My dude, calm down. You need a bump?" Noah held out a stick of gum—probably some upper. Luis waved it away. He was wired enough from the C4 gelato Noah had ordered him at the performance dessert place they'd gone to for dinner. "Look, I did some digging. Kapoor puts on a big show, rakes it in from ads and sponsorships and chat microtransactions, but her actual investments are pretty methodical. She only ends up putting her money into startups she perceives as having real substance. So as long as she perceives us that way, we're golden!"

"Wait—" Luis began, but then a sturdy South Asian man in display goggles opened the door and beckoned him to follow.

"Just stick to the pitch beats and we'll be golden!" Noah called as Luis left. "I'll be in your ear the whole time. Foot on the gas and don't look back!"

The tough escorted Luis down the hall to the auditorium, mocked up with game show flair. The audience—mostly foreigners in town for the COP, though Luis recognized a few faces—was filing back into their seats after the commercial break. In front of the stands, six racially diverse backup dancers in Brazilian streetwear popped the latest filmi moves to a jangling bass beat, camera drones swooping through cartwheeling legs. On the stage, Diya sat in a calfskin leather armchair, wearing noise-canceling headphones and tapping on a tablet. She vibed much more serious than the viral venture scout she'd played stalking the halls of the COP. The production crew waved him forward, and Luis sat. When the music had finished, the lights came on around him. Camera wands lowered back into Diya's luxe interrogation corner.

"So, Blueberry. I hear your founder's name is Noah," Diya said. "'Noah's Ark' is très cute, but is the brand more than a pun? I assume you aren't planning to march two chickens and two giraffes into a big boat."

"The name is aspirational. It's about what we could do if we could really prepare for our climate future," Luis said, visualizing the cue cards he'd memorized. He clickered on their swooping, polished presentation, which began with a Ghibli-styled prophet hammering on a ship. "Biblical Noah was the ultimate inside trader—he got a hot stock tip from God that the flood was coming. He capitalized on that information so well that not only did he survive the deluge, but his investments in genetic capital allowed him to exploit the incredible real estate opportunity of an Earth washed clean of legacy infrastructure and market competitors. His descendants went on to repopulate the planet, which in Biblical terms is pretty much a total win—the ultimate monopoly!"

"There are dozens of climate preparation platforms showing at COP60," Diya said. She leaned forward. "What makes Ark *Diya Kapoor special*?"

"Give her the redevelopment angle!" Noah said in Luis's earbud.

"The UNFCCC has embraced redevelopment as a core principle of moving forward into this new climate era," Luis continued, hitting his stride, "but really we need to be doing pre-redevelopment. Storms only work as economic drivers if you have the equipment and capital standing by to move in as soon as the clouds clear and a special redevelopment zone is declared. That's why neverstorms are so devastating! Ark helps firms and states—even individual households—know when and where to lay that groundwork, so they can make the most of climate opportunities when they come."

"That certainly would have been helpful in San Diego," Diya chuckled. "I hear they're *still* arguing over stadium financing, what, four years later?"

"Exactly," Luis laughed. He didn't know anything about the San Diego firestorm, but this felt correct.

"But I'm not a national adaptation corporation, and I don't fund individual infrastructure projects. What good is Ark's platform to me, as someone who plays the markets and lets other people fortify the dikes?"

"Ark goes beyond risk evaluation to calculate meta-risk," Luis said. "We're partnering with top hedge technicians at Short Stack Dot Fund to offer accumulative financial instruments that profit off the deltas in the larger risk landscape. That means the investing public can support projects that may not make sense now, but could make tremendous sense as the climate shifts."

"You know, there's a rumor going around COP60," Diya said. "A rumor that your platform can actually *predict* neverstorms. What else are you cooking up for us?"

"Fuck that's not what I . . ." Noah murmured in his ear. "Fucking Cheeto! I'll have his guerrilla marketing license."

"Uh, boss?" Luis subvocalized back.

"Just deflect. Try 'predicting the never.'"

Luis felt himself sweating a bit under the studio lighting. He hoped the audience and Diya's online viewers were watching his slide deck and not the perspiration spots forming on his Carcel SoAm floss-silk tuxedo shirt.

"There's predicting the storm, and then there's pre-dicting the 'never,'" Luis said. Noah had rewritten this part of their dialogue model at least a dozen times. "Climate events happen every day. What makes a never-storm a neverstorm is the intersection of atmospheric dynamism with infrastructural vulnerability, followed by media narrative. That's what Ark aims to not just predict but prepare for. Our platform combines the best weather data on the predictions markets with global asset resilience maps, comprehensive investment track-ing, and a powerful algorithm to give our users God-like advice on where and how to spend their adaptation budgets."

"You're playing coy with me, Blueberry," Diya said. She sat back, eyes suddenly steely. "Shall we have tea?"

The lights turned harsh. Diya addressed her crowd.

"Shall we have tea!" she called, and the crowd yelled with gladiatorial enthusiasm. "You know what that means . . . sudden death!"

Backup dancers backflipped in, placed a small table between them and set it with two cups of chai. Diya picked up hers and blew on it. Then she set it back down, pulled out a long SafeVape and delicately exhaled three tiny puffs of red fog.

"Different show this time, Blue," Diya said, stowing the vape. "The chat loves you, but our live audience gets the vote tonight, and they swing older, more liquid. The investing public is here to see innovations that change the climate game, samaj rahe hai?"

Luis nodded. He got the feeling that it wasn't entirely his fast-talking charm that had gotten him this far. Diya had sought Ark out based on whatever underground marketing campaign Noah had been running.

"Let me tell you a story," Diya continued. "I was a little younger than you when Cyclone Vayu hit Bombay. I was away at uni, but I rushed back. I saw the streets torn to shreds, my family estate flooded and waterlogged. But the economy was booming, and it was easy to see that Better Bombay could grow from the rubble. So I threw myself into rebuilding and rebranding. What good was uni when hurricanes were too uncertain for even the experts to predict? This was the first time they used the word 'neverstorm.'"

Diya's nails clinked on the china teacup, their chameleon lacquer shifting to mimic the white porcelain. The lights swiveled to bear down on Luis.

"Come clean with us, Blueberry," she said, enticing as a scalpel. "My sources say Ark is bidding for CalmWorld VR's server power. And everyone knows Short Stack Dot Fund has a skunkworks math team working on chaotic markets. All this 'laying the foundations'—is Ark building a platform that can predict neverstorms?"

Luis knew nothing about this. He supposed it was possible that Noah was trying to position Ark to be the front end of something as bleeding edge as Diya was describing, but "fucking Cheeto" didn't sound promising.

He waited for Noah to give him guidance, but the founder's voice didn't come. Did sudden death mean no assistance? He resisted the urge to tap his earpiece. Instead, he sipped his tea—milky, spicy, curdlingly sweet. He knew what Diya wanted, what the audience wanted: surety.

All the scenarios for full adaptation and climate wealth assumed that *eventually* modelers would be able to calculate the full atmospheric impact of the greenhouse shift, and the world could then smoothly plan accordingly. It had been the holy grail of every compsci firm's R&D lab ever since the '20s, when evidence that chaotic systems might be computable after all won the Wolfram-Fields Medal three years in a row. At every COP there was a rumor that someone had cracked it, and every year that hype dissolved into nothing.

Luis understood the desperate hope. Tropical Depression Mitsubishi had been pinballing around the mid-Atlantic all week, and everyone was on tenterhooks wondering which section of coast would get Russian roulette lucky. No one talked about it. Everyone kept their eyes on the highway, foot on the gas, not looking up, afraid of the sky.

"Is Ark going to predict neverstorms?" Diya repeated. Luis had no clue what to say. Then he heard the buzz of an open line in his earpiece.

"Tell her yes," Noah said.

"Yes," Luis said.

First Saturday

"Shots for my booth boy!" Noah shouted above the music. "Breakfast shots for the best booth boy on the COP floor!"

The waitperson came over, swaying androgynously in a way Luis would usually find appealing. But by now he was too tired and endorphin-swamped to care.

"What can I get you?" they asked. "Glacier melt vodka? Crop failure wheatgrass? Nearly extinct espresso? Our group special this hour is the mass suicide sunrise—Guyanese tequila and orange Kool-Aid."

"Wheatgrass," Luis said, before Noah could order more alcohol. He was determined to sober up and head home. Morning rush hour was long over. Even on Ark's p-card, Luis blanched to think of his parking charges.

Noah and Luis had gone from Diya's aftershow party straight to the midnight opening of the twenty-four-hour CAN'T Party. Noah insisted the CAN'T Party was a COP tradition. "You gotta come, pay your respects, get utterly torched," he'd said. "For the good of, I dunno, the last exquisite coral reef penguin or whatever."

For decades the environmental activist constituency had thrown the biggest party of the conference, mostly for their own community to blow off steam amid the tightly wound disappointments that came with pushing for climate action that ran against the jet stream of market forces. When the world hit 600 ppm atmospheric carbon, they rebranded the event and announced that anyone with a COP badge could come—on the condition that delegates, vendors, and investors all submit to regular guilt-tripping by the servers and DJs. From the drink names to the refugee camp dance montages, the CAN'T Party steered relentlessly into the theme of climate despair. It was a morose affair, but with its massive scale, top-shelf open bar, aggressive marketing, and decadent legacy, it had simply outcompeted all attempts by the UNFCCC to replace it with a more future-positive event.

This year the CAN'T Party had taken over the sprawling, multiclub disco campus at Punta Carrasco, looking out on the Río de la Plata. Next door was the little downtown airport that most of the bigwigs at the COP shuttled into—some apparently coming straight to the CAN'T Party to burn off their jet lag before diving into the higher-level dealmaking of week two.

But for now the party was hitting a lull, and, after hours of dancing and singing, Noah and Luis sat slumped in air couches, watching the eastern light give shape to quick-moving clouds. After winning a tense, countdown-thumping vote at the end of Diya's show, there seemed nothing to do but celebrate. But now, sipping his bitter wheatgrass, Luis felt unease return.

"Boss, did we . . . lie about the neverstorm thing?"

Noah yawned. "I dunno, dude. Did Nikola Tesla lie when he promised to broadcast electricity wirelessly across the whole planet?"

"Yes? Is this a trick question?"

"Well, we'll never know, because J. P. Shithead Morgan pulled his investments before we could find out."

"Are we still talking about the car guy?" Luis asked, confused.

"Point is, anything is possible with the right funding." Noah waved his arms expansively. "Ark just landed a new series and a ton of publicity. We'll staff up and get people working on the 'neverstorm thing' right away. And they'll probably come up with something great, because we'll hire the best! Anyway, Ark's brand is about *preparing* for climate events, not just predicting them. What good is knowing a neverstorm will hit if you aren't set up to take advantage of it?"

"You could evacuate people? Save lives?"

"Sure, but from a market perspective that's small-time thinking, my dude."

This was a good point, Luis had to admit. He dragged himself out of the couch, tested his balance. He was tired-unsteady, no longer drunk-unsteady.

"Desayuno?" Luis suggested.

Noah stumbling, Luis half-carrying him, they checked out of the party and walked until they hit a Hardee's Secret Treasure Cart, which had the best sushi breakfast burritos of any street food franchise. Or at least it felt like that, after twelve hours of subsisting on liquids and room temperature hors d'oeuvres. Then Luis prodded Noah back to his hotel and fetched his car from a glowering valet boy.

It was afternoon when Luis's Mazda finally puffed through aching city traffic. Everyone was headed inland. Luis rolled down the window; no tang of exhaust today, just sopping salt air, rushing by. He plugged in his battery-dead phone. A minute later it awoke with a charming jingle and played the advertisements of the day. Then notifications began to roll in. Luis had a triple-perfect rating from his insurance company, which unlocked all sorts of bonus rewards and discounts from participating Enterprise Alliance Against Distracted Driving partners. He kept his eyes on the road.

He was ramping onto the highway when the rain started, sudden as a whip. He turned his wipers on low, then up to the super-fast speed you had to swipe your credit card to activate, because it wore out the blades. He swiped. Still the rain ran river-thick on the windshield. He felt the push of the wind on his car, the tiny overcorrect of the autopilot. Out the side window, Luis saw smog-capture towers flickering with logos twisted by overclocked spinning.

Anxiety and adrenaline began to pique through Luis's exhaustion. He turned on the COP livestream. Chatter filled the Mazda's surround sound.

"The storm will certainly disrupt some of our programming." Luis recognized the voice of the COP president. "But COP60 will carry on! A neverstorm like this is a perfect reminder of the importance of the great products and platforms we showcase each year. Our hope is that our vendors will step up and offer accelerated and discounted recovery services to our amazing hosts in Buenos Aires."

"And what's your preference on a name?" a half-familiar Weather Channel anchor said. "Mitsubishi bought publicity rights to the tropical depression. Now they're polling their subscribers on which of their 2055 turbo models to name the hurricane after."

"Well, I've always been a TETRA man myself," the president chuckled.

"Shit. Mierda." Luis put both his hands on the wheel. Traffic was still flowing fast, porteños eager to outrun the storm. To keep the fear down, he thought of more languages to curse in. "Chikushō. Merda. Alqarf. Kak."

Then a gust sent him into the next lane. A Porsche behind him honked, piercing the livestream and the roar of the rain. Luis stepped on the gas and swerved back. For a sick, thrilling instant his car seemed to no longer be driving but floating—and then he was spinning, out of control, clenching for some crunch. A full 360 degrees turned around him, a panorama, unnaturally slow. For a second Luis saw the storm sky over downtown: a mass of purple and green, impossible fairyland colors.

Then he was stopped, half in the shoulder, perpendicular to the road, horns blaring around him. All he wanted was to be out of his car. Instead, heart racing, he straightened the Mazda, edged back to a crawl. There was an off-ramp ahead. He took it.

Hands shaking, he looked up storm shelters on Yelp. A few blocks away was a parking garage: three dollar signs, two and a half stars.

Other cities, Luis knew, took their storm shelters very seriously. Even on short notice they provided catering, memory foam cots, bottle service. But Buenos Aires didn't get hurricanes, so their shelters were half-assed. The parking garage had no attendant, just a vending machine with Pepsi and beef jerky, maybe stale water rations locked in a closet. Luis pulled in, found a compact spot, then sat very still for a while.

From his Mazda Luis watched the rain sheeting horizontally over the streets and buildings and into the half-open parking structure. He could hear the wind, a dull whine, like a child failing to whistle, vibrating his car. Leaves whirled through, and a few branches. Hours seemed to pass, though maybe it was just minutes.

All Luis's life, summer in Buenos Aires had been getting wetter and wetter. But this was something else. The water seemed to form fist-dense masses, smacking against trees, street signs, micro-billboards. Occasionally one of these would rip off and spin through the air like a shuriken.

Two men broke his trance. He couldn't tell if they were garage maintenance or other car owners. They shuffled by, toward a rectangular section of the slatted outer wall that had come loose and was rhythmically banging against a Mercedes, which added its car alarm to the noise. The men held their arms over their faces as they pushed into the storm, tried to swing the wall back into place. Then they were sprawling on the floor, bowled over by a palm frond, which tumbled across the garage, a manic paper airplane.

Luis got out of his car. The wind was a banshee howl. Instantly rain, skittering across the concrete floor, soaked through his favorite MeSockaSuSocka brand socks, which he'd gotten on sale at a Fretsy Dot Buy retail therapy spa day. *Weird thing to remember*, he thought. He rushed toward the men, helped them up. Together they threw their weight on the slats. The wind stung his cheeks, pushed against the slats like a sail. He could feel the great weight of the neverstorm moving toward and over him. They heaved and jammed the wall back into place.

Puffing, without words, the three retreated to the interior of the garage, huddled by the vending machines. The screens on the vending machines were dark, that smudged and empty gray; the power was out, Luis realized. He pulled out his phone: no signal from the city grid. Luis peeled off his Berluti suit jacket, shook the water out a bit, tried not to think about the cleaning fee the rental place would charge him.

"Qué quilombo!" one of the men said. "I was supposed to paint my house today."

"Last I heard, they said it would go like this for the rest of the day," the other said. "Can they get the signal back up? I have cam appointments with East Asian clients all evening."

Luis wasn't sure if the small talk was a surreal joke, or the most real thing he'd experienced since his car spun out. The others looked at him, his turn to speak.

"I just secured a seven-figure superseed for a company that pretends to predict this shit," Luis said. The other men exchanged a glance.

"What the fuck," one said.

Second Sunday

The streets were puddled but, thankfully, not flooded. When Luis was just a boy, the Keurig Dr Pepper X-Games Clayco Adaptation Group had extended the Maldonado Creek flood tunnels into a massive underground drainage reservoir slash extreme sports venue. There was, allegedly, no amount of water the city wasn't prepared for. Luis said a vague prayer that the vertboarders and zipline-tag competitors had gotten out in time.

The wind was a different story. Up and down each block, storefront windows had been shattered by flying objects. Glass crunched under his feet as he trudged the couple of miles home. There was too much debris, too many downed trees and power lines to get his Mazda through the streets.

When Luis got to his family's apartment, all he wanted to do was pop an Ambien Extra and sleep. He'd tried to doze off curled up in the backseat of his car, but the storm had been too loud, too frightening, too fascinating. So he'd stayed up for the second night in a row, sometimes crouched with the other two men by the vending machines, sometimes just staring out at the roiling darkness.

Rain had poured through his broken bedroom window, leaving his bed soggy. His posters had been shredded. His collectable plush Bobble-Feet figurines of twentieth-century media personalities with notable feet lay strewn on the floor in their mint-perfect protective cubes. Marcela bustled around, sweeping up leaves and glass, chattering about riding out the storm in the bathtub, which had been scary at the time, but now she felt like she had really accessed a new level of intensity to bring to the one-woman show she was composing.

"Mother and Father were sooo tragique comique all night," Marcela said. "Now Father has gone to restart the bots at some Vaca Muerta shale site, and Mother has gone to check on Abuela Constanza, which should really be your job, but you were off at your COP, your little game show, which, how did that go, by the by? You look awful."

Marcela was often manic, but even groggy Luis could tell that this was not a good kind of manic. His sister was edging on shock, flipping between her several languages to keep her mind off the wreckage. He pulled her into a firm hug, felt her shake a little. Then she hugged him back harder, and he realized he must be shaking too. Eventually she pulled away and declared, in her best impersonation of their mother, that he was no good to anyone in this state. She made him eat a steak empanada while she scrounged up dry sheets and pulled out the sofabed.

He didn't even remember lying down, but next thing he knew his phone was buzzing on the pillow next to him—an expensive level of attention access.

"My dude," Noah said when Luis finally answered. "We're fucked."

"What happened?" Luis had a splitting headache but also felt somewhat restored. "Is it the Kapoor deal?"

"Not over the phone!" Noah hissed. "I'm coming to you."

Ninety minutes later Noah was pacing the small living room. He'd brought an extra-large sausage pizza, on Domino's proprietary, protein-rich, postdisaster Recovery Crust.

"Wild out there, my dude," Noah said. "I mean the storm is whatever, but all the redev firms here for the COP are mobilizing, making a big show of getting their hands dirty. They're out recruiting like mad for the cleanup, arguing over territory, the works. Real *Firefighter Gangs of New York* drama—classic!"

"Wow, I *love* his energy right now," Marcela said in Lunfardo Spanish from the kitchen. She came in and perched on the corner of the still-unfolded sofa bed, picking up a pizza slice the size of her face. Luis hissed at her to go away, but she held up one finger: "I must observe him."

Luis sighed, switched to English. "Boss, what happened with the deal?"

"So she came to my hotel room right after the storm this morning. Diya Kapoor, I mean. And at first I'm like, hell yeah, perfect way to work off that bad weather funk. But she's all business. She wants to put together an accelerated launch plan, a whole PR campaign, all based around how Ark can stop the next Hurricane Mitsubishi Xpander XLRS neverstorm from happening again. So I gotta talk her down, manage expectations a bit. All part of the job, right? But she's insistent. And it turns out she was just playing with me anyway, because then," Noah actually loses his cool a bit, "she motherfucking brings in fucking goddamn Cheeto! That Belgian kid I hired to get our buzz numbers up. He looks like he's been roughed up by Diya's goons, and of course now he's saying that *I* told him to make up that neverstorm shit, which I definitely didn't. So anyway, long story short the deal is off, and we're fucked."

Luis slumped. He looked around the mess of his family's apartment. Marcela had pushed the glass and dirt into little piles in the corners, clearing a path. Now she sat taking copious notes, fingers typing hard on a disposable notetablet. Last time he'd seen her this wound up, she'd gone on a two-day Adderall4Artistes bender, ended up in the hospital.

"Thanks for telling me," he said to Noah. "We've got a lot to clean up, as you can see. Maybe we can touch base on Monday? Talk about getting back to the booth?"

"Back to the booth?!" Noah exclaimed. "No booth boy of mine is going back to the booth! That doesn't make sense. Point is, we're fucked, but we aren't *fucked*, you know what I mean? Diya is out, but her show gave us a ton of free publicity, really attached our brand to the whole neverstorm zeitgeist. This is the perfect time to pivot! Capitalize on this name recognition windfall! How would you like to be the Provisional Chief Operating Officer of Ark Recovery Catering and Festivals? Think of the acronym!"

"Arcf?" Luis tried.

"ARK-AF! Classic, right? The next level. We'll back-burner the planning platform and build up our rep by putting on a massive music festival at the end of the COP. It'll keep media attention on Buenos Aires, rally redevelopment finance from all the UNFCCC firms in town, get people here and abroad excited about exactly the kind of climate-shock-as-economic-defibrillator mindset we've always talked about! I know some of the people who organize the CAN'T Party are interested in diversifying their aesthetic, and they *just* threw a huge event, which they had to raincheck halfway through. They've got all the city's best vendors in their contacts, and all those vendors probably just had their calendars shaken up by post-storm cancellations. It's perfect! Just like *Fyre Festival III: Rise of the LavaLords*."

Getting to be COO, even provisionally, would be a huge step up for Luis, a big get for his CV. But he felt kind of pissed at Noah.

"You know, I almost died last night," Luis said. "We're trying to figure out if our grandparents and cousins are okay. I don't know if I've got the bandwidth to pivot until things get back to normal."

"Come on, dude, this is normal! That's the point. And anyhow helping people have some fun will help them get back to normal even faster." Noah must have sensed Luis's continuing hesitation, because he turned to Marcela.

"You—hip, neo-boheme young person. Who would the teens like to see play this kind of festival? Want to be in charge of the act list?"

Marcela's eyes lit up. "How do you feel about tragic, entrancing thawpop?"

"Yes, I love that!" Noah and Marcela touched their pizza slices together like a fist bump.

Luis knew there was no getting out now.

"Okay, let's do it," he said.

"Hell yeah!" Noah beamed. "Okay, we need a new slogan. How do you like, 'ARK-AF: Make It Rain Like the Rain Don't Stop'? You know what, that sucks. We'll workshop it. Just soak it in."

Second Monday, Tuesday, Wednesday, Thursday

In business school, Luis had flash-audited a course on high-end event planning and had come to the conclusion that it was an activity best left to those hyperorganized people who communed with project management algorithms like New Neo seeing The MetaMatrix—which was not him. Nevertheless, he threw himself into the breach, stringing together a sketch of what *NeverFest: Buenos Aires* might look like, if—hypothetically—they were to put on such an event in several days' time.

By Tuesday the sketch had become a skeleton, and he had holds on a number of key party necessities, to be heli-dropped and snap deployed Saturday morning by some of the greenest vendors shilling at COP. He procured a stage guaranteed to minimize sound and light pollution; high-end camera drones to record the concert as an immersive VR experience (distribution agreements TBD pending paperwork by the various acts Marcela was booking); tents serving certified organic beer, soda, maté, and three varieties of Asian fusion chorizo emp-bao-nadas; and five hundred composting portapotties.

"Boss, how are we paying for all this?" he asked.

"Tell them to send the bills to me, everyone knows I'm good for it," Noah said. "Or at least, most people don't know the Kapoor deal fell through."

"Okay, but," Luis grasped for words to articulate his sense of impending doom, "how are we *paying* for all this?"

"You got to spend money to spend money, my dude," Noah said, and he rushed off to his third lunch meeting of the day.

For all his starving bullshit artist bluster, Noah did have a talent for schmoozing people into attaching their names to his wild schemes. Soon there were serious brands listed on their partner page, and these snowballed into more, with firms and acts calling Luis, desperate to get involved. Everyone assumed someone else had bothered to run the due diligence.

"We're too fast to fail, baby!" Noah crooned when Luis again brought up money concerns. "That's how the whole world works, my dude. Have you seen the latest global carbon debt numbers? Everyone runs one project in the red. If we're a hit, then the funding we get for the next NeverFest will pay this one down. And it's not like neverstorms are going to stop coming, so we'll be fine. Remember, if you die in debt, you win."

It occurred to Luis that Noah's hustle was exactly the kind of something-from-nothing business presti-

digitation that he had always admired. He'd wanted just
this kind of rise to riches: fast-talking, big-dreaming,
judo-flipping every failure into miraculous success. It
was the global capitalist dream!

But as Luis rushed about the festival preparations, he
also watched his city try to recover from the hurricane.
He felt guilty not to be helping with the cleanup, and
he felt guilty for feeling guilty, given that everyone he
talked to was totally excited about his new promotion.
Noah pointed out that the money NeverFest would bring
into the city would go much further than another pair of
hands. Still, there seemed so much to do. Buenos Aires
groaned with the din of twenty-four-hour construction:
beeping earthmovers, concrete clanging into dumpsters,
brawling repair crews fighting over the best-insured
houses. The air was choked with fumes from the gas gen-
erators most buildings had resorted to when power lines
went down, adding to the cacophony. And it was hot.

The pundits said everything was going swimming-
ly, but two of his cousins were in the hospital, missing
work. His parents had moved his infirm grandmother
into Luis's bedroom, which still had flapping plastic over
the windows; a Finnish glass startup had won an exclu-
sive recovery contract with the province, and anyone
who didn't sign up for a lifetime glass subscription got
bumped to the back of the queue.

Kicked out of his room, he crashed with Noah, which
just meant more time partying and working and less
time helping his family get their home and lives back in
order. No one blamed him much. "Business is the most
important thing, borrego," his father said. "This seems

good for you, and good for your sister, too. I'm proud of you for getting a foot in the door for her."

Marcela had hardly needed his foot. Her own had smashed the lock in with one swift kick. She too was camped out in Noah's suite, spending all day and night making calls, sending messages on social platforms Luis had barely heard of, bidding for talent on bookings markets. She begged for favors and called in others, all to secure the right mix of bands, DJs, live artists, and VR auteurs. Her email signature now read "Marcela Soto, ARK-AF Creative Director" and directed press requests to an agent in Toronto.

"Brother, take my p-card and buy Noah new shirts for the next few days," Marcela told Luis on Thursday. "I don't trust the hotel runners to choose something fab, and I can't have him dressing up that Gucci QR-plaid microflannel again to meet our headliner. That is not the life I want to live."

"Who's our headliner?" Luis asked. He'd been checked out on everything but core logistics. Marcela did a smug little dance move and took a bow.

"An absolute legend," she said.

Second Friday

When SAGA arrived the next morning via private share-jet, she didn't look like a world-historical pop icon. She looked like a tired woman who'd been on a plane for fifteen hours. But still, there was something about her, some gravitas beneath the fame, the ur-charisma that had gotten her discovered in the Nordic music hyper-boom of the '30s. For all the famous and powerful people Luis had seen in passing at COP, it was meeting SAGA in the flesh that really made him feel like he was in the big time.

SAGA wore some kind of shapeless linen jumpsuit that Luis couldn't identify, with a straw sunhat so atemporal in cut that it could have passed without comment almost anywhere in the world for the last five hundred years. She looked recognizably like the teen idol that had dominated charts and kids' channels all through Luis's childhood, but also like the gossip-scarred twentysomething whose brief polyamorous marriage to a triad of decadent tech moguls had sent tabloids and financial analysts into an eighteen-month frenzy; and like the inscrutable video artist who emerged from seclusion only to release epic short films that helped define the "xenoearth" aesthetic. It had been years since she'd gone on tour, but her occasional albumlets still drew a lot of attention. Commentators were forever debating if SAGA was slowly easing into de facto retirement or was secretly at the height of her powers.

The story of how Marcela got a line to SAGA and coaxed her down from Sweden would become the stuff of legend in the BA creative scene. But the day before NeverFest, Luis just knew that somehow this weirdly mythical being was gracing his hacked-together relief concert with her presence. Strange how time had compressed and exploded and compressed and exploded again in the two-week rush since he first saw the little blue Ark booth.

"Welcome to Buenos Aires, Saga—sorry, SAGA." Marcela had coached him on how to properly pronounce the all-caps in SAGA's name. "We've got a sun tank prepped if you'd like to shed your jet lag before the show tomorrow."

"I'd rather see the damage," SAGA said.

So instead of riding with her in the luxury shuttle to the Four Seasons, Luis drove SAGA around in his Mazda. She didn't seem fazed by his car, other than to inexplicably touch her toe to the tailpipe—a sad, delicate gesture. They stayed off the highway, took side streets, slow. She spent the ride with her face pressed against the glass, staring, sometimes craning her neck to follow some detail he hadn't seen. He tried to keep his eyes on the road.

"Your first neverstorm?" he asked, dumbly.

"No," she answered. "Sweden had many, when I was young. Thor's Year, they called it—terrible name. And I often visit aftermaths, once I won't be in the way. I hate that it's an inspiration, but sometimes you have to look at a terrible thing to see everything else, you know?"

"A shame you weren't able to see it before the storm," Luis said. "I feel like all storm-hit places must look the same."

"Yes, it's the aesthetic of contemporary globalization. You find it everywhere, like Starbucks."

"And yet, people still travel for it. Disaster tourism is booming!"

"Guilt-tripping," SAGA scoffed. "We'd rather think deep thoughts about our burning world than do anything to put out the fire. The storm here didn't have to happen, but there was money to be made pretending industry had no consequences. And now there's money to be made cleaning up the consequences. On it goes, always leveraging, managing, capitalizing. Never just stopping, even when people die."

SAGA sighed then. "Sorry," she said. "You have been kind to fly me in. I shouldn't be so rude."

Luis realized she thought he'd be offended. Of course, she had every reason to assume he was an arch climate capitalist. He pulled over.

"In the storm, I spun out in the rain. Up there, on that highway," he said, pointing. "Almost crashed. Could have died. And that night, I hated cars. But after, I didn't start taking the bus. I just . . . kept driving."

SAGA gave his shoulder a pat. "When we go fast, we become heavy with inertia."

He continued the tour, chewing on this. Eventually he suggested they get her to the hotel, and SAGA didn't protest. He dropped her off, and she breezed past Noah in the lobby. All this time, Noah had been waiting, in his nice new shirt, for a now-vanished chance to meet the star.

"Wow, she sure is something," Noah said, transfixed. When the elevator doors shut, however, he rounded on Luis.

"Whatever voodoo your bipolar sister did to get SAGA here, we gotta capitalize on it pronto," Noah fumed. "If you'd brought her back on time, I might've been able to incept our pitch before she went to nap in that sun tank. No more weird tourist jaunts. We've got forty-eight hours of potential facetime to have conversations about licensing deals and merch lines, heart-to-hearts about the long-term artistic vision of NeverFest. SAGA could be the face of a new movement in disaster entertainment and finance—*our* new movement. But only if we lock her down with a contract before she zips back to her fortress of solitude. We good?"

"We're good," Luis lied. "I was buttering her up the whole ride."

Second Saturday

It wasn't that Luis felt protective of SAGA—he was a nobody, and she had decades of experience fending off worse predators than Noah. It was more that he wanted to see what she'd do with NeverFest without Noah's . . . what? Interference? He wasn't sure.

So, all Friday through the start of NeverFest—while overseeing last-minute crises and keeping his sister steadyish—Luis did what he could to steer Noah away from SAGA. He made excuses, promised he was working her over. Noah was too distracted to push back much, caught up pumping ticket prices on the scalping markets and tracking numbers from the pledge drive. Or maybe Noah just trusted him.

Even the night before, they had worried about turn-out. They had been late getting the word out about SAGA, and no one really knew whether the local consumer base would have the time or motivation to show up amid the ongoing cleanup. New shipments came in each day, unlocking big chunks of repair work on this or that modular unit type. And there was lots of work people had to do that didn't get captured in the daily economic productivity reports: care work, trauma detoxing, cultivating renewed ties with family and neighbors.

But the crowds came, perhaps ready for a break from the work, or a celebration, or a chance to mourn one week on. Or maybe there was no disaster big enough to stop people from turning out to gawk at something big and weird. Many COP attendees still in town decided to check it out—the CAN'T Party had been cut short, after all. Also showing were many of the workers who had been arriving in droves from the provinces, from Chile and Bolivia, from Brazil and even across the ocean from Lagos and Cape Town.

Mostly they came to see SAGA. Crowds milled around as the first acts played, getting a little drunk or stoned, swaying to the music—but also staking out spots close to the stage. When evening arrived, the loose knots of fans packed in. And SAGA came to the stage.

Her costume draped around her like a robe of wide ribbons. Projections found each cloth panel, mapping her with a tableau, cut to pieces. From offstage, Luis tried to make out the scene on each narrow slice. Here a cloudscape, there a crashing wave, or a seed germinating in time-lapse.

She played the classics first, the fizzy, folky pop hits from her teen years that everyone knew from commercials and movie soundtracks. She strutted around the stage, hyped up the crowd. But there were small changes: codas switched to minor key, lyrics tweaked a tad darker. Luis thought he saw the crowd shiver uneasily.

Then she asked if they wanted to hear new shit, and everyone roared. Her whole, fun demeanor changed. She took to a keyboard and began to pound out an epic, anthemic dirge. Luis watched, rooted where he stood, his hands going numb clutching his obligatory clipboard. It took him a minute to realize it was a cover.

"Last night was shaking and pretty loud," SAGA sang, slow, each word wavering, drawn out, that sinister thawpop hiss—but still rising with a thumping rhythm. Behind her the stage screens showed a sequence of faces, smiling and singing along. As he watched, the faces began to squint and grimace as water and wind pelted their eyes. SAGA began to riff off the half-familiar lyrics, spitting out stream-of-consciousness poetics that dipped through several languages. Like her costume tableau, the words told a shattered story. Some shards were familiar—evac jargon and advertising jingles—others alien, describing Earths that weren't their own, volcanic and elemental, all of them doomed.

The crowd loved SAGA's thawpop virtuosity. They cheered and wept, climbed on each other's shoulders, hurled themselves at the stage. Luis felt a sense of uncorking, masks coming off, young people and old showing their deep anger and anxiety at the world that had been chosen for them. And SAGA, encouraging and amplifying that release with each note of her powerful voice.

Noah came up behind him then.

"She's such a Valkyrie," Noah said, close enough for Luis to hear over the music. "Don't you just feel like we were destined to bring her here for this?"

"Maybe," Luis said.

"I think this is the start of a beautiful business part-nership. What do you say? Will you be my pitch boy again? You worked your magic with Diya. Not your fault how that turned out. Can you do it again?"

SAGA was singing the chorus, shaking and sobbing with affect: "Here I am . . . ROCK you like a hurri-caaaaaaannnne!"

Luis looked at Noah. For the first time, he saw how fidgety and wound up the man was.

"I don't want to be like you," Luis said. "Actually, I wish you didn't have to be like you. We worked so hard for this show, and somehow we pulled it off. We should enjoy it! But here you are, just thinking about the next deal, the next score."

"The fuck is that supposed to mean?" Noah said. He got up in Luis's face, breath chemical with supplements. "I'm giving you a chance here, you little shit. You want a nice car, nice clothes? This is how you get it. Matter of fact, people *like me* are why there's any of that stuff in this country to begin with. The next deal, the next score—that's how people *like me* got billions out of pov-erty. Now you're not going to do your part?"

There it was: the double standard of this whole game. Everything a calculation, until they need something from you. Luis shook his head. "You'll be fine without me. That's the whole point, right? If I had died in the storm, you would have found another booth boy. If this hadn't worked, you'd have just rebranded. 'Foot on the gas and don't look back,' right? That's what the market wants, I guess, but I'm done. I quit."

Noah started to say something, but Luis just walked out onto the stage. For a moment, his eyes locked with SAGA's. Then he climbed down to join the crowd.

Shared Socioeconomic Pathway 4
A Road Divided—Inequality

A STORM FOR SOME

First Thursday

Saga awoke in poverty lighting. She was on her back, head spinning with lost time. She tried to focus on the dim fluorescing strips stapled to the unfamiliar ceiling, bathing her in exactly as many partial-spectrum lumens as international regulations required, and no more. Minimum-cost, minimum-wattage lighting: migraine inducing, cause of millions of cases of blindness around the world—or so the humanitarian groups complained. Saga knew poverty lighting well. She had helped write the sustainable provision policies that made sure countless low-priority homes barely scratched the planet's energy budget. She'd hoped she would never have to see that light again.

When the ceiling straightened and her nausea receded, Saga sat up, checked for injury. She shivered uncontrollably and had a splitting headache. Her body throbbed with bruises she didn't remember, but nothing seemed broken or punctured.

She still wore her black-on-gray negotiator outfit, a post-Euro trouser suit that conveyed her middling position with its too-perfect, machine-measured fit; the truly rich preferred the slight slub of human hand stitching. Her shoes and phone were gone, but the orange COP badge that marked her as a surrogate for a nonstate party was still clipped to her waistband. Someone had covered it in Faraday tape. She peeled the metallic, fibrous stuff away, waved the badge over her head on the off chance there was hardware nearby to ping the passchip. The badge had some basic display functions, but no GPS, no messaging. It was useless outside the conference venue—and this, she was starting to realize, was not the conference venue.

The room was small, more like a cell. She sat on blankets, the scratchy kind that got dumped on refugee camps at the end of the luxe-down recycling chain. The walls were dirty panels of pressed wastewood, snapped together, smelling of glue. One panel looked like a door, hinges on the outside. She crawled off the blanket, shoved her weight against it, felt the wood bend a bit, heard the clink of a heavy chain. She doubted she could break it. Wastewood was cheap, but it was tough, too.

In one corner was a bucket covered with a thick rubber flap. She peeled back the flap, and the stench of feces in chemicals poured into the room. It was filled with the gray mix of sand and waste-dissolving agents that Anglophones called "pour-a-potty" and everyone else called some translation of "shit sludge." The bucket was small, just big enough for her to squat over—if she aimed.

One thing in the room didn't fit the slum-in-a-box aesthetic: a black cone of camera, wedged into a ceiling corner.

"Fy," Saga swore. "Skit."

She began to sift and gather her memories from the chemical haze that fuzzed the morning. She assumed it was still the same day, but she had no idea of the hour. She didn't think she was dehydrated enough for much time to have passed. Of course, her captors could have given her a saline drip while keeping her unconscious for as long as they wanted. In the silence and solitude of her cell, it was easy for her mind to run wild.

She remembered being followed. She'd left the venue compound. Something had delayed her morning session on EDA-9—fresh concerns about powers and mandates for the new global carbon bureaucracy's Bureau of Urbanization. The UN wanted to bring the cosmopolitan cores under a unified mitigation regime, and that meant fiddling with a lot of dispensation requests for individual development projects. Each core wanted to expand, but doing so without expanding the pool of citizens with travel papers would lead to instability in the real estate markets. Her patron had wanted to take a more direct hand in the minutiae, and so had excused Saga for a few hours.

"Go out, my dear. Get some morning air," Diya had said. "We'll be at this until after late-day tea at least."

So freed, she'd left the gated zone and walked around Old Buenos Aires, thinking she'd stray just a bit outside the arcology and find some greasy food. Saga didn't like traveling, but her job made it necessary. So she had been trying to develop an appreciation for the advantages of globe-trotting. The streets had seemed pleasant enough, all quaint shops and polite performers. A passive middle-class landscape maintained by philanthropy and patronage, edged with odd pockets of semi-stable lower-class life. She'd eaten a butter-sweet Argentine breakfast and headed back.

But on the way back she'd been followed. Saga had seen the van first: a scratched-up, white delivery box, but it didn't drive like it was automated. She kept catching it in the corner of her eye, always a block away, no matter how she zigged or zagged. Then there were the boys, two of them, dressed like homeless kids. She remembered wondering how long they'd be allowed to wander before security ran them off. They too had moved a bit too fast, without that swaying urchin shuffle. She'd quickened her pace, cut into an alley she thought too narrow for the van to follow—a mistake, she now realized—and then she'd tripped. Before she could get up, a weight had kicked into her back. A hand had grabbed her hair and turned her head to the side. A cloth had been pressed to her face, the quick smell of cleaning chemicals. Chloroform? She supposed the old ways were sometimes still the best. Next thing she knew she was waking up in this waste-wood room.

In her finishing school, during the UN's VIP liaison training programmes, all through her life in fact, she had heard about the scientific studies that claimed that what really set the global elite apart was not IQ or willpower but metacognition—the practice of thinking about one's thoughts and emotions. Thus she too tried to cultivate such reflection on her own life. So Saga took stock of her feelings.

She felt scared, obviously. This was rational since, unless she was much mistaken, she'd been drugged, kidnapped, and was now held captive. She felt angry, since she'd worked very hard to attain a position where her personal autonomy was respected by nearly all the COP's power players. She felt frustrated to be snatched away from work she cared about. She felt a second, deeper fear, the kind that rose out of pure sense memory. The hardness of the floor, the textures of the walls, the claustrophobia of the room—all evoked childhood traumas and anxieties, her years clawing her way out of the camps. And beneath all that, she felt an eagerness, a sense of opportunity, for in any black swan there was a chance to prove herself, to outmaneuver her rivals, to turn danger to her advantage.

Saga stood, still woozy, and addressed the camera.

"Can I please have some water? I'm very thirsty."

Then she repeated herself in Spanish. Even without her phone, she was pretty fluent. You didn't keep a non-nepot job at the COP by relying on translator apps. She thought about trying Portuguese, just to be thorough, but didn't want to offend some touchy local sensibility.

She had chosen to ask for water because it seemed like the hardest request to refuse. You can't keep a human

locked up for very long without a plan to give them water. Whatever happened next would tell her something
about her situation, and she could assess from there.

Saga sat on the blankets, stretching, occasionally
standing to repeat her question to the camera, sometimes rapping politely on the door. Soon she began to
get thirsty for real.

"One of the Kapoor negotiators didn't show this afternoon." Noah daubed hand-gathered wild caviar onto a
cracker of organic ancient grains. First rule of working
for the UN: always raid the parties' catering. "The Swedish one, Saga Lindgren, on EDA-9."

"Isn't that the one you have a crush on?" his American
contact Marta asked. She was between sessions, taking
calls while getting a full-body massage at the Virgin
Group pavilion. Noah could hear the squelch of oil on
flesh in the background. "Never mind, I don't want to
know. Where is she?"

"Parties took more informal work time in the morning. System has her leaving the venue via the south
checkpoint. She used her per diem card at a cafe. Never
showed back up."

"You talk to security?"

"Who do you think told me? Surely you don't think
I spend my days here stalking every random surrogate
who turns me down for a dinner date?"

"You really want me to answer that?"

"My guy knew we had special interests in EDA-9,
pinged me the report."

"You gonna tell me why I should care about this, Campbell? I'd be a lot more forgiving of your attitude if I had a reason to think you weren't wasting my time."

"Apologies, ma'am. What I'm getting at is, EDA-9 is stalled."

"Stalled? They couldn't just get a substitute?"

"The Kapoors wouldn't have it. Claimed the item was too delicate to discuss formally without the proper staff. Seemed like bullshit, but you know Diya Kapoor. Picky. Saud backed her, because, you know, Saud loves to stall. The Walton and Bezos teams pretended to stay out of it but made noises like they agreed. Probably had quorum to force a continue anyway, but it would've been a bad look. So sec-gen tabled it, rather than risk giving offense. The Kapoors have the facilitator on the take too, if I had a guess."

"Remember when you could bribe an asset, and they'd actually feel some loyalty to you? Maybe they'd double-cross you, or triple-cross on a bad day, but they wouldn't just let the next bloke in line take a slice of the same pie. It's like we're all buying timeshares."

"Peace through a web of overlapping interdependencies. That's civilization, right? Now the parties are all in little clumps, cubing about something."

Marta barked something away from her phone, presumably scolding the masseuse for some indelicate touch. When her voice returned, it was even more annoyed than usual.

"Is this a real concern or thin cover for more Deccan real estate nonsense?"

"I really couldn't comment, ma'am."

"EDA-9 needs to get in consensus this year. We've got projects in Texas waiting on those resettlement mandates. You tell the Walton and Bezos negotiators I said their interests, as Americans, are for a swift resolution on all matters EDA-9. They're still American citizens, right?"

"Barely. What about Diya Kapoor?"

"Is this your way of asking for permission to track down your little Swede? Worried she's run off with some handsome Argentine peasant before you could wear her down?"

"Just trying to cover all the bases, ma'am. If someone got Lindgren to play hooky to kick off this palace intrigue, it would be good to know who."

"Fine. You're the fixer. Find out what happened to her. I'll give you State clearances, just let the Kapoors know I'm the one having you help them out. Diya will owe us a favor if we're the ones to return her favorite pet."

Saga had never been kidnapped before, but she had been through the UN's mandatory social hazard training. Being on the diplomatic circuit made one a target, and, ironically, courtiers like her got snatched more often than the princelings, politicians, and meritocrats, all of whom traveled with personal security. She had little power beyond representing the calculated positions of her patron, but replacing her would be enough hassle that the powers-that-be—either the Kapoors or the UN— would pay out a reasonable sum for her return.

Kidnapping was all very businesslike, her training had explained. A cottage industry made possible by block-

chain-encoded escrow transactions that neither kidnapper nor ransom-payer could cheat on. Most kidnappers
were men of means—Saga assumed they were men, since
she doubted many women would be so cavalier about
violating someone's bodily autonomy for profit. They
had legitimate jobs that gave them access to gated cosmopolitan cores. Snatching her was probably just an
opportunist side-hustle. Soon enough the arrangements
would be made and she'd be dumped with another headache close to an arcology border.

But the person who opened the door did not fit this
narrative.

First, there was a rap on the door, and a young, male
voice told Saga, in Spanish, to sit at the other end of the
cell. The camera was on her, so she didn't bother trying
to disobey. She heard a click, watched the door open
fifteen centimeters before jangling taut the chain that
secured it.

"Still thirsty?" the voice asked, in decent English.
"Water-man came late today."

A water bottle skidded into the room. Saga grabbed
and thirstily opened the bottle, which felt slimy in her
hand. It was made of organic, biodegradable plastic,
probably the type that turned into porous gel within a
few days to prevent relief shipments from being resold
on the black market. Saga drank fast, in case it fell apart
in her hand.

The water-man. So, Saga was in a place without clean,
consistent plumbing—her cell most likely part of a real
slum, not just the facsimile of one as she'd half suspected.
Saga knew what it was like to live where safe water was a

scarce commodity, where getting a drink meant walking a few klicks, standing in line, waiting for an armored vendor cart to wheel into your camp section. As she finished the bottle, she shifted to peer through the gap.

Her captor was lithe and ragged. He sat regarding her, straight-backed and cross-legged a meter away from the door—just out of reach if she were to try to lunge for him. He had no mask covering his wide face. His clothes weren't tactical, and he had none of the businessman's machismo that she'd been expecting. A rubber band slung his hair into a dirty bun. A pistol was tucked into his waistline. Saga guessed his age at late-twenties, but who knew these days. He was probably younger. The rich, with desenescence medicine and gene-tailored cosmetics, could look like teenage supermodels well into middle age. The poor, meanwhile, could grow old very fast indeed.

A minute passed, then another.

"Sorry. Out of practice," Saga finally said in English. "I used to be very good at being thirsty. Best to save it, yes? Quench it slowly? Can't be sure when the water-man will come again."

Saga rolled the empty bottle back. The man retrieved it, but still didn't speak.

Going into any negotiation, the strategy immediately forks. The first path is to make it about the other party: stay quiet, let them talk, explain themselves. Most people with any kind of power love hearing their own voice and will tell you everything you want to know about them. So listen, figure out what they want, really want, and find a way to give it to them—either as an exchange or by helping them see that fulfilling *your* desire will really be fulfilling *their* desire. This was Saga's preferred strategy, one that had helped her claw her way out of the austerity ghettos, into schooling and position.

But there was another negotiation path, which was to make it about you. Do all the talking, overload your interlocutor. Batter them with psychological violence, if you have leverage, or just with the sheer force of your need. Make your story, your personality, your existence so vivid that it eclipses their own. Eventually, if you do this right, the other party will simply bend to your will. They will give you whatever you want and tell themselves it's out of sympathy or fear or whatever reason. But really it's because you have made yourself the most important person in the room, and people naturally tend to do whatever the most important person demands.

Saga had seen both strategies work countless times at the COP, despite the elaborate diplomatic structures designed to contain and soften the power of that core ego-driven mechanism, which on the international stage was called "hegemony." The trick, of course, was that whichever strategy you chose, it was hard to know whether your adversary was in fact plying you with the other.

Waiting was getting her nowhere, however, so Saga decided to probe the second path.

"In the twenties my family lost our flat in the EU housing crash. So, it's funny, this stuff is familiar." Saga patted the 'fugee blankets, half an invitation. She tried to force pathos into her voice and posture. "Then we had Thor's Year, that flutter of the Gulf Stream, the North Sea superstorms. They moved us into camps, in Gothenburg. That's where cholera came back. You know cholera?"

Still no response, but something in his expression changed, a resentment softening to mere skepticism.

"You had to learn to love your thirst that year, nurse and nurture it, let it play out like a numb ache, like hunger," Saga said. "Rather than some acute and disorienting need."

The man pulled out another bottle and slid it into the room.

"Save this one," he said. "But don't let it melt."

This, she decided, was her cue to change tacks.

"My name is Saga. Saga Lindgren. I'm from Sweden." They'd already seen her name badge.

"Are you from Stockholm?"

"Gothenburg. On the other side of Sweden."

"Too bad," he grinned. "It would have been funny."

Saga blinked. He was joking about Stockholm Syndrome, she realized. She knew the term. It referred to when a prisoner formed a bond with their captor. Usually it implied a kind of brainwashing, being turned to the ideology of the terrorists. Or else it implied the hostage falling in love with the dashing criminal, fear and tight quarters unlocking forbidden fetishes. Clichéd movie nonsense.

In recent years, however, the term had gained another meaning. Social critics had coined "Stockholm philanthropy" to refer to the type of neofeudal fealty that foundations and private aid groups cultivated among the under-tended populations they took responsibility for.

Provision, propaganda, and recruitment promises tempered the backlash against austerity and travel controls.

Saga examined the way the man watched her, looking for signs that he was seeking romance—or sexual violence if he didn't get it. Saga had learned early in life that being able to sense the intentions of men was necessary to her survival and advancement, and she was good at it. She looked for flicks of the eyes, flush in the temples, an exuding of will into her personal space. She found none, and relaxed just a bit.

So, a political joke then. Maybe this wasn't about money, but some kind of grievance. Saga sensed that he wanted something from her. He needed her approval, perhaps her cooperation, and was trying to figure out how best to get it. He was joking about Stockholm Syndrome because he knew this made him vulnerable.

"You don't look like a professional kidnapper," she said.

"You know what that's supposed to look like, mina?" the man said. "Maybe I'm just a hobbyist. Do it for the fun."

Saga wanted to ask if it was fun drugging her, dragging her off the street. But this was all a kind of negotiation, and that wasn't going to get her what she wanted.

"What can I call you?" she asked instead.

He eyed her, perhaps weighing the power dynamics just as she was, but then shrugged, a lopsided movement. "Call me Luis."

"So, hobbyist Luis, is this a catch-and-release kidnapping?"

If he was going to joke about her situation, she needed to as well, otherwise she was conceding power to him.

"Depends," Luis said.

"Hobbyist or not, I can get you money. My employers will pay for my safe return. I can show you how to contact them securely. We could get this done today, make it easy on everyone."

"What if I don't want money?"

Saga scoffed. "Everyone wants money."

"Sure, sure," Luis grinned again. "How much money would I need to buy a vote in that big meeting of yours?"

Saga blinked. She had not expected such a question.

"Depends," she said.

He cocked his head. "Okay, mina, explain it to me."

This could be dangerous territory, but it was also a chance to prove her value and willingness to cooperate, earn trust. Trust, she knew, could be leveraged to give her options—options that might include escape.

"Do you know the term 'voting share'?" she began. Luis didn't reply either way, so she explained. "In a company, maybe many people own stock, but only those who own a significant percentage of the stock have enough of a share to really vote on how the company is run or get a seat on the board. The COP is much like that. The players at the COP are those who command enough of the world's resources to have a say. That means states, conglomerates, a few families. The COP seeks to form proportional consensus between these players about what is to be done with the planet."

"So?"

"So wealth doesn't *buy* votes. Votes *represent* wealth, because wealth means power to get things done. Used to be only governments had votes. But the result was an inability to craft policies that could actually be implemented, because they didn't get buy-in from those who actually held real resources. So the Davos Agreement redefined who got to be a party to the negotiations and how much voting share they got."

"Sounds like you aren't worth that much, huh?" He waved away her objection before she could speak. "It's okay, mina, I don't want money. I want you to help me stop this."

Luis held up a pamphlet, once glossy, now crinkled and smudged from having been passed through many hands. The cover showed the sleek glass curve that Saga recognized as the Buenos Aires city arcology: a contained, urban environment, recycling water, filtering air, generating much of its own energy and food, designed to be impervious to various disasters. The opening ceremonies had included a promotional video celebrating it as one of Argentina's great climate achievements. The compound included much of downtown, bringing already-developed commercial districts and major attractions into a single, managed habitat, but the image on the pamphlet mostly showed the actual arcology, built over ten years with UN-blessed investment. The structure covered the Costanera Sur Ecological Reserve, crossed the Río Dique, and replaced the old Retiro-Mitre railyards with dense, sustainable housing.

Upon opening the previous year, the cozy condos had briefly been a hot collector's item among the world's luxury real estate connoisseurs. Argentina had pushed aggressively to host the Sixtieth COP, both because they wanted the prestige of bringing the conference back fifty years after Buenos Aires had hosted COP10, and because they hoped to squeeze a new round of sales and profitable trades out of wining and dining the visiting climate elite. Indeed, the empty units were being used to house many of the COP attendees—including Saga—as well as some of the conference activities. To see a picture of the place where she had been sleeping probably twelve hours earlier, now clutched hatefully in Luis's hand, was to Saga the biggest surprise since she'd woken up.

"Here." Luis stabbed a finger at a spot on the rendering. "Villa 31. One hundred and twenty years old, and always they are threatening to move us out. Always we stay. They put up their walls and glass next to us, but we stay."

"Villa miseria" was a local term for slum, Saga knew, the Argentine version of the Brazilian favela, but on a smaller scale. Villa 31 must be where she was—a famous, intractable settlement, just yards from the COP venue.

He threw the pamphlet into the cell. Saga picked it up. It was a policy briefing trifold, pitched at COP attendees, detailing the host city's plans to expand their arcology. In the fine print was a list of dispensations the UNFCCC could provide that would speed up development: loans, World Future Heritage consideration, human rights waivers to resettle squatter populations. She'd been reading many of the same talking points earlier that day.

"They want to push us farther. Got to make room for new walls," Luis said. "I've read our government's latest proposal to the UN. It's all public records. Legacy democracy, right? Any other year, there'd be nothing I could do. But *this* year, I get lucky. UN comes to town, to right on the other side of the fence. I got a list of everyone involved in EDA-9, put the word out, just in case. And then I get lucky again. I get you. So I figure, maybe you and I do a little negotiating ourselves, comprende?"

"I see," Saga said carefully. "I understand. But you know I'm just a surrogate, a representative. I can't decide policy."

"What about your owner? What can they do?"

Saga bristled. "I don't have an owner."

"Come on." Luis said it like he couldn't believe anyone could be so stupid. "Everyone is owned by someone, even in your world."

There *was* a bit of machismo there, Saga decided. Or maybe just bravado painting over nervous tension. He wasn't entirely wrong, of course, but Saga realized that her predicament was more complicated than a normal kidnapping. Money was easy to get, but policy change was a much bigger ask. The timeline could get tricky. There was over a week left of the COP. Would Luis hold her all that time, waiting for the outcome he wanted? Or longer, to ensure there was no backtracking? And even if Saga got Diya to spend the necessary political capital to squash the Argentine arcology's progress in EDA-9, there was no guarantee it would end the development project in Buenos Aires. Perhaps her best bet was to convince Luis to free her and trust her to carry out his agenda on her own.

But first she needed to pinpoint the exact register of Luis's grievances, if she was going to get leverage over her situation.

"'My world'?" she stalled.

"Your world. The plutes' world. The core. The arcologies. The part of our city they put behind gates and
high flood walls."

"I'm not from your city."

He shrugged again. "Doesn't matter. Same in every
city, right?"

He was right, but still she prodded. "How would you
know? You ever been?"

"You think we don't read books here? Watch shows?
Ever read *The Scattered Fortress*? *Archipelago of Power*?
Everyone knows what your world is like. Most of you
just don't know ours."

He was wrong about that, Saga thought, at least in her
case. But here she sensed an opportunity.

"Okay," Saga said. "Why don't you show me?"

First Friday

They fed her rice and beans and gray nutrient slurry.
This was the standard meal of the global aid-indentured.
In the late '40s, a fad for food printing had promised to
end world hunger. Red Cross, UNICEF, Amazon Smile
Response, and others had rolled into poor countries and
slum communities with shipping containers full of meal
replacement powder and just-add-water fabricators for
printing the stuff into patties and kabobs. The fabbers
quickly broke, or were stolen, stripped for parts, and
the manufacturers stopped making the compressed flavor beads that supposedly turned the chalky nutrient
substrate into an approximation of chorizo or banana
pudding or whatever else. But the powder had been

delivered in absurd quantities, and so it lingered on for years, found its way into local cooking practices. People mixed it with curry, slopped it onto flatbread—a goopy protein shake that made everything taste worse. But it was food, technically, and as long as you had a barrel or two sitting around, the accountants at the World Food Programme considered you well fed, delivered no more help. The barrels had tracking software that snitched if the powder got dumped. So the aid-dependent ate the stuff: reluctantly, resentfully, constantly. Now Saga ate it too.

When the sun went down, the fluorescing strips dimmed off as well—maybe they didn't have batteries to back up their solar, or maybe the grid just didn't waste off-peak watts on Villa 31. Saga dragged her pile of blankets into a corner and curled up in a sort of nest. It had been years since she'd slept this rough, away from sensor-adjusted mattresses in diplomatic circuit hotels, plush four-posters in her patron's guest bedrooms, sur-round-screened sleep pods when she flew business class. Saga was grimly proud to discover that these comforts had not made her soft. She did not toss or turn, but fell into the aggressive slumber of her youth: wary, ready to wake and move in an instant, but also stock still, deter-mined to rest.

For the rest of Thursday and Friday, Saga tried to make sense of her captor's life. The door would open wide enough to slide in water or food, and through that fifteen-centimeter window she saw glimpses of a busy room. It was crowded with mismatched furniture—too much of it for the space, as though they'd cleared out a

real bedroom to make her empty cell. Luis didn't seem
to mind her watching, and when she made no moves to
break out he started leaving the door chained but ajar.

Luis kept their conversation going a few minutes at
a time as the hours unfolded. He asked questions about
the workings of the COP, which she answered truthfully,
mostly, unsure what other information sources he had.
She was also unsure still if it was safer for her to be useful
or useless. She asked him about his background, and he
would tell her absurd and contradictory stories about
life on the streets of the villas miserias, his dashing rise
through the ranks of the gangs. He made the politics of
Argentine slum life sound every bit as complicated as
UN negotiations, and she could believe that, though she
mostly didn't believe him on the particulars.

He came and went, tugged away by messengers or
texts to his phone. The latter was a tortured-looking
device, clearly hacked together from parts pulled out
of several different locked-down, surveilled "'fugee
phones"—cheap mobiles distributed to disaster and
migration zones by the big tech conglomerates in an
attempt to get displaced and out-market populations
into one walled garden platform or the other.

Others passed through the room outside Saga's door
as well: two ragged-eared boys, maybe the ones who
snatched her from the street; three grown men who
moved as a pack and glared at Saga; an old, withered
woman who seemed not to speak; several small children
delivering notes or running errands; a teenage girl who
shared Luis's wide face. Often these others sought Luis's
advice in hushed, head-bowed conferences—lieutenants

reporting to their don. When they arrived, Saga saw Luis's swagger grow more exaggerated. Maybe that was weakness, but maybe it was leadership.

As she sat and stretched and listened, she tried to determine how much of their discussion was about her—or if Luis was juggling multiple responsibilities. Clearly not all of Luis's compatriots were comfortable with the risks of holding a foreign national hostage. Some of them covered their faces with bandanas or air filtration masks or full-head rag wraps that warded off even the most advanced facial recognition. From the rise and fall of their whispers, the words that trickled through her door, Saga sussed that three basic positions were being debated.

Some wanted to ransom her back to the core as soon as possible and had been doing research on the blockchain mechanisms to get that done. Some—like Luis—wanted to see if she could be useful as a contact to the core, and were trying to devise ways to open up negotiations across the almost epistemic gulf of inequality. And a few felt that keeping her or ransoming her were both too dangerous, and they'd be better off getting rid of her, then devoting their efforts to disappearing and covering their tracks before the powers-that-be came knocking on their wastewood doors.

Occasionally Saga would ask for things—mostly just to see what would happen. Now that Luis had acknowledged that he needed something from her, she was testing the bounds of their relationship. She asked for a shower, and he brought her a bowl of wet rags and a stack of small, individually wrapped hotel soaps—the same kind that had been in her room in the core. She wondered if he had connections to the arcology cleaners. As the activity waned on Friday, she asked for a book. Luis gave her a stained, stapled printout of *An Unneeded Class*, a radical satirical pamphlet that proposed the elimination of "the unnecessariat," while being coy about just which group that term referred to: the structurally unemployed or the leisure-living rich.

"You really from a camp?" Luis asked her Friday evening.

"Six years," Saga nodded. The maybe-sister was looking at her curiously from a ways behind Luis. She wore an oversized charity-cotton tee, which had been spray-painted with a graffiti stencil that made a rainbow-colored giraffe. Saga noticed the distinctive Bill & Melinda Gates Foundation ampersand between the giraffe's legs. "I was a little younger than her when I left."

The girl picked up on Saga's attention and came closer, sat down and hissed excitedly in Luis's ear. Luis scowled, but said to Saga, "What was it like?"

"At first it wasn't so bad," Saga said, deciding that with the girl present this was an opportunity to appeal for goodwill. "Lots of people making the best of a hard situation. We'd moved a lot since losing our flat, living with relatives, so the camp seemed like just another stop on that trip. And they tried to make us comfortable. There were all these little touches of—you know 'hygge'? *Comodidad*. Not much room, but warmth, bright design, for a while. I don't know. Maybe I was too young to notice what was shit, or people protected me."

She flashed a significant look at Luis's sister, then continued.

"But the storms kept coming, and the people kept coming too. Things got crowded. And even that, I think, could have been okay. But outside the camp, something changed. The country stopped feeling sorry for us. We were Swedes, their neighbors, but they stopped seeing us that way. We became alien. Lazy leeches. Then disease came, and all their fears were confirmed. The austerity became asperity—harshness. The camps became punishment, you know? For being unclean and dangerous. It was a shock to my parents, I think, losing their sense of country."

"They still there?"

"The camps? Not exactly."

"I mean your parents."

Saga shook her head. "Cholera. Where are yours?"

Luis glanced at his sister, but neither of them answered. "Why'd you leave?" he asked instead.

"Why?" Saga was baffled by the question. "Who wouldn't want out? I hustled, made myself valuable to those in charge, found openings to get myself noticed. A visiting dignitary saw my potential. She sponsored me to go to a real school. It was a very lucky chance."

"Did you ever go back?"

"By the time I finished my education, there was not much to go back to. Eventually people left, or stayed and were absorbed into the periphery. There had been a belief that the refugee system would get people back on their feet when the crisis ended. Give them real housing, compensation. But eventually people realized that wasn't coming, and they found their own ways to stabilize, somehow. Just much poorer than they had been before Thor's Year."

"What about those people that protected you? You do anything for them?"

Saga felt herself being drawn into a trap here. In trying to earn sympathy with her story, she had played on Luis's ideological turf. Now he was rounding on her.

"I don't feel guilty about surviving," she said, trying to take the accusation head-on. "Or about taking my shot to get a new life. I didn't have a choice. Everyone would have done what I did, if they could. Now I have put myself in a position to make things better for people all around the world."

"You mean now you work for the plutes," Luis said. "That's the problem, mina. When you serve just yourself, you end up serving them."

"You think I should have done what you did? Start a gang, run the slum, start kidnapping anyone just a bit better off? What's your alternative?"

Luis gave his lopsided shrug. "My alternative is solidarity. Until you have that, everything you do will be just 'surviving.'"

"I have solidarity with the future," Saga bristled again. "We are close to getting the runaway train of civilization under control. The rationing and investment since Davos has swerved emissions dramatically, and in another COP or two, we may have the political tools to begin deep afforestation and sequestration. That means a better, more stable future for both your world and mine. What else do you think the COP is for?"

"I think everyone in your better future is going to wonder why my world didn't fuck up your world when we had the chance—before the walls went up. I know

I do. Forty years of the powers-that-be doing nothing, and people just kept behaving themselves. Where was the ecoterrorism? The sabotage of pipelines and power plants? The bombs at petrobanks? The COP held hostage at gunpoint?"

Saga took a moment to digest this. It was, oddly, a sentiment she occasionally heard whispered by COP attendees as well, drinking champagne or scotch after hours, when long negotiations had worn down everyone's sense of propriety.

"I don't think that would have helped," she said eventually. "You may not like it, but real climate action began when elites decided future generations mattered more than contemporary people who were already beyond saving. Now, if we are very lucky, our grandchildren won't have to go through what I did—what we did. When the storm passes, we can work on solidarity."

"It's only a storm for some, mina," Luis said. "You're still talking like you know us, like we're the same. But if you walk up to those big walls with that suit and that badge, they're gonna let you in. I don't think we're going through the same thing."

"Why don't we walk in together? You know I'm sympathetic to your complaints. We'll talk to my patron, see what she can do. I'll take you to an EDA-9 session. I'll yield my time for you to make your case."

For a moment, Luis seemed to actually consider it.

"Yesterday, you said I should show you," he said eventually. "Tomorrow, you can see our place. Maybe you will see why I don't want to leave, to just survive alone. Then we can talk about making our case."

Noah didn't like the way Diya Kapoor regarded him. Like he was an urban songbird and might go from protected to pest in an instant if he shat on the wrong bureaucrat's car. She was better than most of her peers; unlike a lot of back-stabbing scions, egotistical-but-insecure nouveau riche, and judge-y, cultish meritocrats, Diya had inherited responsibility for her family's Bollywood fortune with a real sense of noblesse oblige. She even seemed keenly interested in climate science and so came to the COP in person instead of sending her professional surrogates to negotiate on her behalf. But no matter how enlightened the plutes got, it was still human instinct to shoot the messenger.

"As you know, madam," Noah said, "the UN is very invested in reaching proportional consensus on EDA-9 this COP. As is the United States, our coordinating stakeholder on the item. So when we heard that you were having staffing difficulties on this item, I took it upon myself to look into the matter."

"I hope you don't think me silly for wanting to wait for my dear Saga," Diya said. She dipped an exquisitely engraved biscuit in her tea. "She and I have had in-depth discussions about my reservations and expectations on this item, and I simply do not trust anyone else to represent my position here."

"Of course. I understand, madam. However, I am afraid I bring unfortunate news."

Diya likely already knew the gist of what he had to say, of course—that's why she'd been sour to him since they'd sat down—but the plutes had a Victorian need for manners and small talk. Performance and ritual were

everything in this world, so they plodded through anyway.

"Dear me, what has happened?" she replied, fingertips
to sari-wrapped collarbone.

"It appears our Miss Lindgren has fallen afoul of
some local ruffians. Gangs from the informal settlements
sometimes operate in the margins of the economic transition zone around the venue. Miss Lindgren wandered
a bit far afield and encountered them. We don't have
video or imagery of the incident, but we have loosed
algorithms provided by the American FBI, the Argentine
Federal Police, and bots curated by some of Google's
highest testing teams, as a special favor from the Digital
Conglomerates Group. All agree that her disappearance
corresponds to the movements of a pair of vagabonds
seen prowling the edge of the priority perimeter and that
of a delivery vehicle we now believe to be hackjacked."

"That's just awful! Saga is such a loyal surrogate. Tell
me you know where she is."

"The vehicle exited the bounds of the security platform and then may have lit out to areas unknown or even
scrambled its designation and reentered at a different
point. As I'm sure you appreciate, effective surveillance
is difficult to maintain in times of economic discontent.
Buenos Aires is no different than Mumbai or Los Angeles in this respect. The system is rife with blind spots,
darknet markets for tracking exploits, dispensation
sprawl."

"Hmph. I often wonder if our contractors and personal security don't prefer it that way. Such an imperfect
system keeps them useful, after all."

"I couldn't speculate, madam." Noah had noticed that there was nothing those with power liked to talk about more than the untrustworthiness of those they relied upon—or each other.

"My team is, of course, looking into this as well. We are keen to buy her back, but as yet have received no word from the kidnappers. I trust you will let us know if the ransom notice comes to the UN instead."

"Indeed, madam. We have briefed the appropriate personnel to monitor spam filters across all likely channels. And I will continue our investigations and keep you updated on the situation as it unfolds. However, I must also ask whether, even in these extraordinary and sad circumstances, we could work with you on accommodations to get EDA-9 back on schedule."

Diya didn't answer him. Instead, she waved to a servant, who had been standing near-invisibly in the corner of her suite. The man wore a white, buttoned smock and a turban, to match the reclaimed-colonial aesthetics in style among the South Asian aristocracy. The servant presented a tablet, on which Diya scribbled something. Then he left the suite. Seconds later, an identically dressed man appeared and took his place in the corner.

"Do you know my dear Saga, Mr. Campbell?" Diya asked, resuming the conversation on a new topic more to her liking.

"Only a little, professionally."

"Perhaps I shouldn't be telling you this," Diya said, which Noah took to mean that he was about to be sucked into some twisted bit of plute drama. "But the truth is I have a deep personal investment in our dear

Saga. I have known her for some years. I was on a relief tour of the Nordic tragedies, just a few years after I came into my responsibilities. You would not believe the sad state of affairs. The cultural loss! The human waste! It was a great refutation of those who claimed middle-class rule could thrive even in times of disaster. But there were glimmers in the refuse. In Sweden I met an orphan girl who had nothing, but who had managed, in the midst of the cholera, to make herself useful as an assistant to the camp warden. She was so sharp, so attentive. I lifted her out of that place, as I have with many others who showed promise in dire circumstances. But dear Saga remains one of my favorites, for once she was out of the ghettos she needed so little of my help. In fact, as her career progressed, it was she who mostly helped me, particularly here, where the stakes are so high."

"Thank you for sharing this context, madam. We recognize, of course, the vital role that strong personal relationships play in these negotiations. We never could have made so much progress cutting emissions if—"

"You miss my point, Mr. Campbell. I have worked hard to fend off other patrons who have tried to sink their claws into my dear Saga. As you know, she's an extraordinary woman." Diya pursed her lips and gave Noah the tiniest of glares. "But given the cascade of security errors and inadequacies I'm being asked to believe in to make sense of this supposedly random kidnapping, I wonder if there isn't a simpler explanation: that one of the other parties is behind this devilment, launched as an attack on myself and my interests."

Noah took this in. "Madam, I understand your concerns. I assure you that we have no evidence that points in such a nefarious direction. Even in our age of careful cuts and hedges, bad luck can strike at any of us. But we must carry on with the work in front of us."

"Then carry it on without my consensus. If EDA-9 matters to you, I suggest you find a way to return my surrogate to me."

First Saturday

On Saturday morning Saga awoke to rooster crows buzzing through the sound-deadening wastewood walls. She stretched, composed her clothes, waited. Luis had promised a tour of the area he wanted protected from core expansion. This was either her chance to make a break for it or her chance to prove her sympathies and begin to negotiate for her release.

When the door opened and, for the first time, the chain was taken off, it became clear that she'd need to wait for the right moment to make a bolting escape, if that was her plan. Luis wasn't there, but his lieutenants crowded the outer room, guarded the doors. During lunch they joked a little, speaking Lunfardo slang she only half followed, but they were obvious about keeping an eye on her. Then a knock came, and they hustled her out of the building.

Saga only had a moment to look around. After two days in a windowless room, her gaze went up and she glimpsed in one direction the looming, shiny bulk of the core arcology, close enough to cast a narrow noon shadow down on them. In the other direction, over the tops of much lower buildings, was a scraggle of rusty cranes and, in the distance, an ominous cloudbank sky. Then calloused hands ducked her into the white, boxy van she'd been snatched by, and she was back in a tight space with Luis.

"Today we see Villa 31," Luis said from his seat on a bench. The girl who might be his sister sat beside him, directing the vehicle with a scratched-up touchscreen.

The van was a hijacked automated delivery cube, and those didn't have windows for thieves to break in through. Instead the inside had been jugaaded up with screens: cloth, spray-painted silver, had been pinned to the sides, and a disposable projector wand had been glued to the ceiling. The projector had splotches of dead pixels, and sprayed everything with a tiled "Nouveau Noblis Luxury Raves" watermark. As the van started moving, the windows came on.

They were rolling through a mess of colorful squares. Squares of brick and breezeblocks, of corrugated steel and illegal plastics, of wastewood and plywood and broken PV panels. Leaks were patched by poly tarps. Everything bristled with rebar. There were other shapes too: trapezoidal lean-tos and skinny rectangular apartments that had been jammed between buildings. But the ends of shipping containers were squares about 2.5 meters wide, and those were everywhere Saga looked—a standard form that regularized little patches of the geometric chaos.

Technically the scenes projected as windows could be footage from anywhere, or even fully fictional renders.

But Saga doubted it. From the way the images moved as the van jolted and jostled, Saga could tell they probably came from cheap patches of camera pasted on the sides. Saga watched the villa scroll by, trying to focus beyond the shapes and colors to the details and the people moving within.

The denizens of the villa miseria looked nothing like the "favela chic" models found in high-fashion magazines. Nor did they look like the skeletal figures whose photos had driven so much charity in the twentieth century: the famine-shrunk, the mine-mutilated, the gas-scarred. Their immiseration was neither glamorous nor grotesque. They looked mundane: people going about their days, walking to jobs, running errands, corralling children, sitting watching the thickening traffic, perhaps in pairs, with a game board propped between their knees. Sex workers draped themselves out of windows, vendors set up little sprawls of knickknack wares, cartoneros picked at rubbish piles. All were less fashionable than the pedestrians you'd see on the streets of Manhattan or High London or Abu Dhabi, but not without style; they were less wired with wearables, too, though not by much. Saga had trained herself to see the way the powerful saw, and through those eyes the people of the villa looked admirably atemporal—as though they might have been plucked from the downtrodden crowds of any century.

But the Saga of the Gothenburg camps, who knew what to look for, saw the small details. The way pedestrians clutched plump satchels and knapsacks, having learned not to leave their valuables unguarded or leave home without a go-bag that would get them through a storm, police sweep, or riot. The way their skin was darker than that of passersby in Old Buenos Aires, not just because they were more tan, more exposed to the hothouse sun, but because of the racial biases inherent in the grand socioeconomic-climatological system that filtered some people into arcologies and some into slums.

She noticed certain items of clothing and accessories repeated throughout the crowd: surplus aid goods that got dumped and woven into villa culture, just like the nutrient slurry. Technocrats had understood for decades that swamping a community with free clothing ruined any chances of building a local textile industry. It had since been deployed quite deliberately wherever upstart growth might be inconvenient. Other objects were missing: nice handbags, jewelry, anything that might have been bartered away or given as bribes on a refugee's trek out of a disaster zone. The dispossessed were not supposed to have nice things, and the whole system conspired to take them away. To make up for the lack of adornment, some wore elaborate hairstyles or dyed their dull charity tees.

Like everything Saga had experienced since she woke up in poverty lighting, the view from the van was half familiar. She'd once heard of something called "the Reverse Anna Karenina Principle": all places that were happy in the climate crisis were happy in their own unique ways, while all unhappy places were the same. At the Gothenburg camp she'd survived in her youth, governance had no doubt been different than in Villa 31—more under the thumb of Euro austerity cops, she assumed—and the milder winters here allowed for a more open, airy kind of sprawl. But the two places shared a similar social and material culture.

The van pulled to a halt amid honking traffic. In front was a clump of electric vehicles, some linked together by braids of cable—trains of engines sharing one good battery in the face of global lithium rationing. Behind them more vehicles joined the jam, blocking their exit. Everyone shuddered forward as they packed in tighter. Sloppily printed bikes and rickety powered scooters wove between the cars, filling in the gaps.

Luis, who'd been narrating to Saga bits of villa trivia, now bent over the tablet with his sister. They poked at a map, trying to parse the incomplete and contradictory traffic data being fed to them by the mesh networks that laced the slum. Saga tried to peer out the "windows," but the projections were flat and moving her head offered no better angle to view the gridlock.

"Is traffic always this bad?" Saga asked.

Luis glanced up. "Lots of porteños trying to get to the core today, to work at your conference or through to other villas. But they closed the gates, so everything is getting blocked up."

"They do that often?"

"Whenever they feel like, mina. You think they explain themselves to us? Maybe it's a busy day at your COP."

"It's Saturday, right? I remember the schedule. Things should actually be quieter today . . ."

Luis gave her a look that might have been appraising but might also have been worried. He nudged his sister, spoke in Spanish. "Marcela, fly up, yeah? Give us a look."

The girl tugged a cardboard box out from under the van bench, pulled out an improvised drone. Little plastic rotors and a half-sphere of camera had been glued to a frame of crossed twigs, tied with twine. With taps on her tablet, she got it whirring, then hopped to the back of the van. Saga tensed, instinctively preparing to bolt, then caught Luis's warning eye. One of his lieutenants opened the door, and Marcela quickly tossed the drone into the air.

The screens lining the van became a disorienting, upside-down panorama. The ceiling was the street—the image harder to make out without projection screens, but obviously showing the traffic, slowly shrinking as the drone rose. Vehicles stretched out in both directions, ending at the high walls, motorists gathered around control boxes arguing with whatever algo or guard controlled access to the core. Around the clumps of traffic, bikes and scooters swarmed through alleys, trying to find paths out of the slum. The drone lifted past the armored, fenced-off highway that ran over and through Villa 31, and they saw that that route was jammed as well.

Then Saga glanced down at the walls and saw the weather.

Thick, black clouds were pouring into the city from over the brown water of the Río de la Plata. They had a weight to them, an oppressive potential. Saga remembered that feeling from when she had lived through Thor's Year as a child. An arctic ice shelf event had created a roiling, swirling knot of temperature and pressure churn that had stalled major jet streams and flung storm after impossible storm into the North Sea. Whatever

was bearing down on Buenos Aires might be warmer—
perhaps the unpredictable tropical phenomenon she'd
heard UN scientists call "neverstorms"—but she knew
the danger was the same.

Saga raised her eyes. Everyone else was still looking
at the ceiling, trying to find a path out of the gridlock.
She put her hand on the wall of the van, felt the vibra-
tion of the rising wind, and waited. When the gust came
and the drone went tumbling, the view from its camera
jolted in a way that would have made her eyes water—
if Saga had been looking. But while the others blinked
and started, she jumped. She burst out of the van before
Luis or his lieutenants could move to stop her. Then she
was half-running, half-contorting, twisting through the
traffic, free.

After two days of captivity, it was bizarre to be moving
unfettered around people. Even as adrenaline and fear
made her heart pound, she wanted to laugh, to shake a
stranger's hand. Instead she kept moving. These villa
denizens might not have anything to do with her kidnapping, but she didn't know how much reach Luis might
have. She couldn't trust them not to give her right back.
After all, they were, as Luis would say, from his world,
not hers.

Instead she headed for the arcology gate.

Saga felt rain on her head. It was pleasant at first—
the day was much hotter than the last few had been.
Then she was soaked. The sharp gusts of wind she had
predicted in the van now blended together into a single
howl that shook the shanty buildings on either side of
the street. A turn of her head confirmed that Luis and his

men were following her, despite the rain. She was racing both her captors and the danger of the oncoming storm.

As she wove through the cars, bare feet splashing into suddenly wet potholes, body shoving against other bodies, all chaotically seeking shelter, pursued by hostile forces, Saga suddenly felt a vivid sense of remembrance about the moment. The childhood familiarity of the slum and the storm mixed together into something deeper and stranger—a déjà vu that didn't fade but intensified with each step she took. She'd been through this before, she knew—in Thor's Year, hot sleet and high walls, and maybe other, forgotten times as well—and somehow she was certain that she'd go through it again.

Saga found the sensation oddly calming. She felt sure that she'd be okay, that she'd hit the arcology gate and burst through into safety, find a dry, calm place to ride out the storm.

She shouldered her way through the crowd, toward the opening in the high walls. Where cars usually entered for inspection and passage, a metal grate had been pulled down, and behind that a sheet of clear polyglass snugged into a rubbery lip in the road. To the right was a revolving door for pedestrians—a big one, on a platform that Saga guessed would spin slowly as entrants were scanned for weapons and sniffed for contraband. Today it was still and sealed.

Saga inched forward, pulling and prying herself through the people yelling at the gate, and she pulled out her orange COP badge. Her hand strained toward the black square she knew would read the diplomatic priority in the badge's chip and let her in—along with whoever else shoved in with her. That was good, she decided. The arcology should offer shelter to these people. But there were so many of them. She seemed to be swimming through wet, scared flesh. Then the badge hit the square.

Nothing happened.

When the rains started, Noah stood atop the wall and watched the slums around the arcology fill up. In reality he was in a lounge, surrounded by screens that shifted with his eye movements and vents that blew a warm, wet, gentle wind, ruffling his hair and flecking his cheek with bits of moisture. For caution's sake, the COP attendees had been asked to evacuate deeper into the venue compound to wait out the unexpectedly strong storm. The bulletproof glass walls of the arcology were strong, but one could never be too careful when VIPs were present. So the curators of the COP art track had hastily set up this viewing room to give badges an opportunity to reflect on the power and majesty of "next nature." Ergonomic cushions were strewn around, and Noah could hear the clink of tea china as others waited their turn in the next room. Still, the effect of the installation was quite immersive.

"Below him," brightly painted roofs of corrugated plastic peeled off single-room shanties, whipped down narrow streets. Rubbish piles stirred, tossed themselves into the air, and then were slurped up by a rushing flood. A storm surge had broken the port levees and was sloshing over the train tracks into the villa. Cars stuck in traffic slid and jostled into each other. The murk, unable to spill into the arcology grounds, now pressed against the high wall from whose simulated vantage Noah watched. The waters had nowhere to go but into the slum, and up.

It was unfortunate, Noah thought. These informal settlements always lacked proper drainage. The social ills that clustered in such places made them hard—and expensive—to retrofit. It was easier to build an arcology that could protect a city's culture, productivity, and ingenuity from the worst storms. But arcologies necessarily had walls, and walls left some people out.

The climate transition required a retreat from the inefficient sprawl that had encouraged unrationed consumption, an end to the cheap energy that had subsidized the lifestyles of millions beyond their productive contributions. Future-proofing every spot where the unhoused pitched their tents and built their shacks was an untenable path. It was a struggle to even establish stakeholders who could meaningfully represent such populations in the climate consensus process. What could be done for people whose lives were so backward that the global platforms couldn't even fill proper datastreams for them?

And yet, Noah couldn't help but notice the artistry in some of those painted shanty roofs torn off by the storm, the ingenuity in the jugaad hoses and wiring that ran from shack to shack—a microgrid of illicit water, energy, maybe even data, now ripped apart and flapping like ribbons in the hurricane winds. He wondered, in a vague way, about the lives of the informals who'd built all that. There were no people visible.

Noah realized that here and there the screens blurred, editing out imagery that was too graphic to show in good taste. The biggest blur undulated at the edge of the wall. He leaned toward the screen and commanded it to zoom in. At that resolution, it was in fact many blurs, pressing together into the watertight gates, perhaps asking to be let in.

Noah found himself wondering about Saga Lindgren, whose disappearance was provoking so much confusion and paranoia. The storm was certainly scuttling the drones he'd dispatched to search for her. Where was she during this sudden catastrophe?

On a whim, Noah opened the display's backend, entered his UN credentials, and puttered through the menus until he found the graphic content filter. He toggled it off.

The blurs resolved into waving arms. Dozens of informals, splashing around, clung to the gates and each other as water rushed past their feet. They looked unremarkable, other than their distress. Except one. Noah zoomed in further. Yellow hair—a bit unusual in the slums, Noah assumed—and a gray suit. Her hand waved an orange square at the camera.

"Shit," Noah said.

The water came in so quickly, Saga almost thought she'd lost time again. Suddenly it was at her ankles, then surging up toward her knees. She swayed with and against the crowd at the wall as the flood tugged them all off balance.

Then guards appeared on the other side of the gate. They looked nervously at the water and the desperate crowd. One of them made eye contact with her, and then spoke into his comms. Behind them another man emerged, wrapped haphazardly in a rain slicker, holding an umbrella. Despite the terror and confusion of the moment, Saga recognized him. It was the UN staffer, the American, who had made a pass at her more than once over her last few years of COP attendance. The American—Campbell—fastidiously shook moisture from his umbrella, then waved at her. Saga pulled herself closer to the metal grate between them.

"Open it!" she shouted, unsure if he could hear her over the din of wind and water and wailing.

"Just you!" Campbell mouthed. She read his lips as much as heard him. He pointed to the side, at the stalled revolving pedestrian door. "We'll bring you in. Just you!"

The other side of the glass was dry. The wall held back the storm surge, and the rain was whisked away by pumps, probably dumped out into the villa, worsening the flood.

Saga turned and saw Luis. For a moment she thought he was still pursuing her, but no. He stood on top of a car, smashing in the windshield with the butt of his pistol. As the water began to carry the vehicle away, he helped a woman climb out of the driver's seat and pointed her toward the swaying shacks that were the only higher ground. Others were getting organized as well, a chain of arms leading people away from the dangerous clusters of shifting vehicles. But the waters kept rising.

"Open the gate!" Saga pleaded. She shook the grate. "You're killing people with this wall!"

Campbell looked at her like she was crazy.

"It's locked down," he shouted. "I couldn't even if I wanted to! Get in the door while there's still time!"

"Send help! Let the water drain into the compound! Do *something*!" Saga felt the heat of tears welling on her wet face.

"Lady, they're not your problem." There was no trace of cruelty on his face, just calculating apathy. "Are you coming in or not?"

Saga looked back at Luis again. For a moment, her eyes locked with his. In that instant she took stock of her feelings. She felt animal terror, and guilt, and anguish. And beneath that, in a shameful pit, was hate and malice, directed not at the flood or even at Luis but at the struggling people around her. Helpless, shapeless sorrow welled in her heart.

She shoved her way toward the revolving pedestrian door. She yanked aside a middle-aged woman who was clinging to the push-bar, heard her splash into the murk. Saga wedged herself into the open quarter section of the door, and a moment later the platform started to turn. It was agonizingly slow. Hands found the edge of the door, but she peeled and scratched them away.

A quarter turn, and she was locked in, water still up to her waist, waiting while the machines patiently scanned her. It was like a tiny room in the middle of the wall, camera wedged into the corner, grim fluorescing strips embedded in the ceiling. She almost laughed, thinking of the cell where she'd spent the last two days, remembering waking up in poverty lighting. She half-fantasized about the room filling up like a magician's box, drowning her. But then the rotation resumed, and a moment later a gap appeared. The water sloshed away, and Saga spilled out into the dry room beyond.

A cashmere blanket wrapped around her shoulders. A hot thermos was pressed into her hands.

She looked back, but the water had risen high enough to obscure the villa.

"You okay?" Campbell asked, offering her his arm. She didn't take it. She wanted to retch.

"I'll survive," Saga said, and she trudged toward the arcology, safe and dry behind the high walls.

Shared Socioeconomic Pathway 3
A Rocky Road—Regional Rivalry

HOT PLANET, DIRTY PEACE

First Saturday

Noon

The hum of the wind jumped an octave, drowning out the staccato of distant gunfire, and Diya knew she'd been right. Even through the smoke and smog, she could smell the storm coming.

Her heart pattered with that familiar adrenaline kick, bringing a titillating edge to her low-level, everyday anxiety. But she made herself stand still and drag away the last millimeters of her rumpled cigarette. When she had mastered herself, she craned her neck out the window to get a glimpse of the clouds she knew were coming up the river. The Río de la Plata felt close enough to spit in, but she couldn't see much through the architectural welter of campus concrete and torn-up drone nets. She supposed she should be glad of that. Anything that could line-of-sight you could also kill you, and so decent security made for a lousy view. She'd have to get closer.

Diya flicked away the pinch of her fag and closed the scratched, blacked-out hallway window. Then she slipped back into the auditorium to catch the end of the conference session and grab Saga.

"—Sadly, they couldn't make it to this COP to argue the point, but we believe there's just not enough good data available on East Asian economic and conflict conditions to correlate oceanic carbon in the South China Sea with expected emissions. But even though we disagree somewhat with their methodology, our own results corroborate the Kaho-McVey study's ultimate findings. We believe global atmospheric CO_2 concentrations are indeed in striking distance of 600 ppm. Unfortunately, with the loss of GeoCarb and the other observatories aboard the International Space Station, we're likely to stay in the dark about just how much radiative forcing that translates to in this unprecedented era. Thank you."

The speaker wrapping up was one of those mousy, ethnically ambiguous, Euro-ish women that border guards everywhere in the world always seemed to let through. If COP53/60 had a type, that was it. Diya had watched the COP's demographics shift year by year. She had even felt her own affect converge toward whatever men with guns and papers found neither threatening nor interesting. It was, she thought, a sign of the times.

The de facto abandonment of the UN and its Framework Convention on Climate Change had brought many such changes to the yearly conference still called "the COP." Though organizers kept the name for nostalgia's stake, gone was the pretense of negotiation or activism. What was left was a wholly academic enterprise—RINGO types, cataloguing the arcane details of the climate collapse as best they could without the historically unique combination of big state budgets and international cooperation.

This got harder every year—as every presentation seemed obligated to remind them. The COP remained a prestigious event with academic institutions sponsoring attendance, but who showed up had become much more about who could actually get to the host city. Assuming they could find a host city that was safe—whatever that meant.

So, a certain cautious, forgettable, under the radar look came in handy for most of the participants. Except Saga of course, blonde and towering in the front row, who got to the COP and the other climate conferences through sheer willpower. Diya often wondered how much that determination came from her military upbringing, and how much was fueled by the misguided hope that the COP could do more than just document the slow, sagging decline of civilization.

Saga headed toward the podium as the other academics in the room started to sway toward the exit. Before she could corner the speaker with questions, however, Diya swooped down and hooked her colleague's arm.

"Saga, my dear," Diya said, voice edged enough that Saga wouldn't protest. "Would you come for a walk with me? You know what they said about women going out alone, and I'm afraid I desperately need more smokes."

Saga reluctantly agreed, and Diya led her out of the building and north toward the river. When they were out of earshot of the Facultad de Ciencias Exactas y Naturales, Diya figured that it was safe to talk.

"I was right about the storm."

Saga blanched. "The tropical depression from the radio? It's going to bring a squall near here?"

"'Near here' might be an understatement. And it might be a little more than a squall."

"You said yourself that your trough-collapse model was an outlier," Saga objected.

"An outlier that happened to be correct."

"How can you be sure?" The sun was still out above them, and the air was smoggy hot. Saga pulled a cloth and plastic mask from her bag and slipped it over her face.

"I'll show you," Diya said. "But first I really do need smokes."

Diya got cigarettes from a little armored kiosk where the Universidad de Buenos Aires's science and tech campus gave way to an old park memorializing the victims of twentieth-century state violence—now ironically a staging area for the halfhearted military training and indoctrination required of all UBA students, and younger men as well. The dozing kid in the kiosk still wore fatigues, dirty from early morning exercises. He had a wide face and a buzz cut. "L. Soto" was lettered in marker on his name patch. Diya woke him with a cautious rap on the poly shutters and a call of "cigarrillos, por favor" into the slot before stepping to the side, just in case the kid was packing.

Saga tried to buy a bottle of water, but unsurprisingly L. Soto refused to take her euros. With the storm coming,

Diya advised her not to spend any of her pesos—either the Argentine ones or the Crypto-Pesos Americano still sloshing around Latin America. Diya thought about explaining to the kid about the hurricane, but her Spanish sucked, and her translation app had been broken by Argentina's aggressive national firewall. Instead she just pointed at the darkening sky and said, "No bueno."

"Do you think they know?" Saga asked, as they walked away.

"Securidad? I suspect it will be hard for them to miss, soon enough. Let's get a better look, before they herd everyone into shelters."

Diya lit up again and led Saga through the memorial, now defaced by rightist graffiti and vandalism from the last time some globalized-nationalist gang had marched through campus. On the other side was an old brick riverwalk, and the Río de la Plata. The walk and piers were pocked and chewed up, probably by mortar fire from across the bay-like river delta during a past spat with Brazilian-occupied Uruguay. Like Sherpas reading moss growth for landslide omens, Diya and Saga eyed the rain-blunted wash of the rubble edge, deciding the damage was old enough for them to approach safely.

At the water they looked out at the clouds roiling in from the Atlantic, midday sunlight silhouetting the mass with a burnt-orange halo. Diya saw hints of purple and green, half-flashes of lightning, that black-gray haze of rain meeting the water. The edge of the clouds was becoming more solid and discrete, coalescing into a single front bearing down on them. Diya knew from experience that storms looked different to each individual. She wondered what Saga saw. To Diya, the hurricane looked like a giant hand, closing into a fist.

Diya watched the storm with conflicting emotions. Studying the bizarre, sudden, overclocked storms of their monstrous age was her life's work, and she couldn't help but feel excited to get so close, to glimpse the results of all that unnatural climate energy with her own eyes. Since Cyclone Vayu had ripped up Mumbai around her at the age of twenty-one, launching her on a grim and daring scientific path, Diya had seen a handful of super-storms in person. But none in the last decade; travel had gotten too restricted to chase storms on short notice. She felt perversely lucky to be here, now.

She also felt ruefully unlucky and deeply scared. Climate vulnerability was on the rise, particularly in areas like South America, where drugs, resource extraction, post-Amazon ecosystem collapse, nasty weather, and nastier nationalism mixed in a deadly cocktail of disaster conflict. It was one thing to know those statistics intellectually, to talk about them with academics at the COP. It was quite another to face the possibility of becoming part of such a statistic herself.

"What a sight." Saga was rocking on her heels. "How long do we have?"

"Until it hits land? An hour, maybe two before it gets bad."

"Did you get a chance to read my presentation notes? I was thinking I may need to adjust the last couple of slides, if there really is consensus forming around the Kaho-McVey timeline."

"Yes, you're very thorough," Diya said, scanning the rest of the murky horizon. Around the Río, smoke from coal plants, or maybe barrel fires, drifted up to join the clouds. Beneath her cigarette, the air tasted of ozone and soot. "But I wouldn't count on getting your session tomorrow."

Saga sighed. "No, I suppose I can't. Everything has to be hard, doesn't it? Even when all we want to do is understand how bad off we are, the world always finds a way to make it difficult."

"Always," Diya said. "Come on, let's find some higher ground."

They made their way back to campus, where security forces were swarming. The kid in the kiosk was gone. They kept to the armored box's shadow and peered at men in uniforms. Some bustled about rumbling vehicles, some stood guard outside buildings. One unloaded boxes from a truck, stopped to take a call, and then loaded them back on again. Like ants when the first raindrops start falling, Diya thought.

"Just Argentine securidad?" Saga speculated. "Though I see some armbands."

Diya pulled up an app on her phone that checked network and radio traffic for competing cryptographies: a geiger counter for potential conflict. "We've got a few layers of fragmentation going. Someone could be making moves."

"Might just be military command shuffling," Saga said. "Or a couple of different New Peronist Front caucuses organizing among the troops."

For a few years now, Diya had been a sort of mentor to Saga. Before joining the climate conference circuit as an emissary of a Nordic archival collective, the Swede had never traveled much, though she'd been adjacent to more than a bit of action in her home country. Diya appreciated that Saga remained interested in the abstruse particulars of every country's individual flavor of nationalism and strife—though Diya herself had long been bored by such details.

Diya shrugged. "Does it matter? Could be any number of random-country-generator agitators, trying to unite this or separate that. A bunch of brainy foreigners with few direct political connections—that's a ransom prize any faction would be tempted by."

"You think they'll use the storm as an excuse to grab and hold us?"

"I don't know, but bad weather always dirties up the situation. I don't want to end up in some besieged neo-Peronist bunker if one of the opposition milicias decides to roll through. Or if anyone else makes moves. The Americans and the Chinese both have stakes in Brazil's occupation across the river. Let's find somewhere civilian to ride out—"

As if to punctuate Diya's point, they heard a string of loud pops. A ways off, a long stretch of campus grass exploded into flying dirt. Everyone around ducked for cover. Diya heard the high-pitched buzz of a drone dopplering by overhead, then more lawn went up, closer this time. The big nets strung up between buildings made it hard for the drone to target anyone inside, but strafing the security forces gathered outside was easier. The locked kiosk provided little cover. Diya grabbed Saga's hand, and they shuffled away.

The riverwalk was more exposed, but outside the area the drone was strafing. They kept under a line of neglected trees until the walk met a road. Then they hugged the walls of the buildings across from the water—old restaurants and offices abandoned for being just outside the range where Argentina's automated defenses, such as they were, could effectively shoot down adventurous Brazilian drones. Diya wondered if the one they were fleeing was from across the water or from some malignant group here in the city: perhaps a drug cartel in need of a distraction, or an antimigrant terror cell mad at the university, or even something like El Movimiento Naxalita Americano taking potshots at securidad as revenge for some clash. Along with the movies her family used to make, India's particular style of guerrilla insurgency was one of her home country's most popular global exports.

A few more times they heard the lethal whine, first receding and then getting closer, joined by another angry buzz. Diya risked a glance up: two quadcopters swooped around each other a hundred feet up, dogfighting. Their movements had the uncanny, jerky precision of two automated systems competing without human intervention. It would have been interesting to watch, if the drones hadn't been spraying bullets at each other, spatting into the ground whenever they aligned vertically.

"Didn't the briefing packet brag that Argentina hasn't had a declared conflict on its soil since the Cold War?" Saga said ruefully as they moved through the brush.

"Probably. But they've certainly had a lot of hot peace," Diya said.

Diya's churidar was ripped on the sleeve, and she was wheezing in the dirty, urban air. There was a gap in the wall, with more tree cover beyond, so they forced their way through some bushes. On the other side they found themselves surrounded by squat concrete structures anachronistically painted with a pattern of stone blocks. The structures had arched windows and domes, beyond which poked the too-green fronds of fake palm trees. Saga, silent and focused all through the attack, started at the sight of a plastic donkey, staring at them.

They had stumbled into some kind of theme park re-creation of an ancient Middle Eastern village. Weird, but for the purposes of avoiding random drone fire, this was pretty much perfect. They ducked inside and wound through connected hallways, passing murals and life-sized dioramas of what Diya assumed were scenes from Catholic mythology. The place looked old but lovingly cared for, freshly painted. Religious fetishism had a powerful appeal in chaotic times. Diya had soured on religion after seeing Hindutva mobs and Muslim refugees from flooded Bangladesh beat each other to death in Kolkata water riots.

Diya feared that they'd need to stay under cover in the faux-Bethlehem until the rain arrived to drive the drones away. This was not the higher ground they wanted. But eventually the gunfire stopped. After a few minutes of not hearing much, they exited the hallways and made their way back out to the street. There was no all-clear siren or text alert. There never was. Things were simply quiet, and they just had to hope quiet meant safe.

As they headed inland, the wind picked up. Diya dug out her cigarettes—smushed but salvageable—and lit up, let the nicotine smooth over the cortisol shakes. Black clouds rushed over them to cover the sky. Raindrops began to plop onto the ground—gentler than bullets, but no less dangerous.

2PM

Bastardo was a hip, funky, faux-dive bar built in the teens. In the '20s it had sprawled up into a multileveled nightclub and restaurant. Then the venue's ambitions had contracted with the economy, and extra space was put to various and sporadic uses: gambling den, electronics repair shop, pirate radio studio, artist live/work commune, cartel front, soldiers' kitchen during a brief Brazilian occupation, safe haven during sieges, halls of government for short-lived autonomous zones. Because of its proximity to campus, rail, highway, and the museum row of the more stately Av. del Libertador, Bastardo had become a neighborhood fixture that packed more history into four decades than any one building had a right to. The bar itself had moved up to the top floor, where it lived long enough to become an actual dive.

When Diya and Saga dodged inside, they knew only that it was the tallest, sturdiest-looking building within a five-block radius. They pushed in as the rain turned to pouring and followed the arrows up several flights of stairs. Their journey since exiting the Holy Land park had been less eventful than the drone violence on campus, but in a way more stressful. When drones are fight-

ing, you just find cover. It's much harder to know how
to handle human encounters when time is short and the
weather turning hostile. Diya had hoped to get further
inland, but military convoys, roadblocks, crowds, and
a pair of car accidents had routed them northwest, no
further from campus than when they had started. Soon
their priority had shifted to simply finding any place that
was high and dry and open to the public.

"Perhaps we should have stayed in the conference
hall," Saga worried, surveying the bottle-littered bar, the
dim lights, the graffiti-painted walls, the clutter of stools
and uneven tables, the row of curtained-off booths. The
room smelled of grease and meat and spilled alcohol. "At
least we would have been with the other participants. If
something went wrong, we'd have strength in numbers."

"I like it," Diya said. She was pleased to see that several other patrons were ignoring the faded, heavily van-
dalized "No Fumar" sign hanging over the bar. "There's
food and drink, and no windows to shatter and let the
rain in. Besides, it looks like we aren't the only ones
from the COP who preferred not to leave our fates to
securidad."

She nodded at a cluster of damp foreigners who had
taken over one corner of the room. A few of them she
recognized from the conference.

Diya and Saga had bugged out of the UBA venue before lunch, but the danger and anxiety of the last two
hours had kept hunger far from their minds. Still, Diya's mantra for bad situations had for decades been "eat
when you can, sleep when you can, piss every opportunity you get." So while Saga clearly wanted to join her

peers in the corner and compare notes about the storm and the drone spat, Diya first dragged her over to the bar.

They ordered pita sandwiches, fries, and weak beer, with tall waters. Diya was impressed that Bastardo served seemingly drinkable water in glasses, rather than plastic bottles, and she persuaded the stoic bartender to also fill up their water flasks for a few of Saga's useless euros. No guarantee the plumbing would still work if the storm brought a flood. Then they went and used the toilet, washed out some of their scrapes and smudges, squeezed their clothes a little dry. Diya heard a single sob escape the stall where Saga was cleaning up.

Diya stared into the marker-scrawled mirror. In her face she saw the lines and pallor of a long day bound to get longer. Behind it, a long conference, a long trip by shipping flotilla from her home in Pune, a long life of bad weather and dirty peace. She thought, for a second, that she looked blurry, out of focus. She reached out, felt the mirror buzz against her fingers. The howl of the wind and rain had become so constant, she'd almost forgotten it was there.

When they came out of the bathroom, their food and beer were ready. They sat at the bar and tucked in.

The food wasn't great. Everything was torturously fried in salt and oil. Trade had tightened up a lot in the last few years, and whenever that happened food culture always took a hit, especially in the Americas. Diya pitied them. Desi cuisine had roots in the subcontinental ecosystems, the spices and crops that had grown there for thousands of years—at least until the weather started changing. European colonialism in the Americas had created inconsistent mashup cuisines that relied too much on access to global markets.

Still she inhaled the meal ravenously. She *had* been hungry, turned out. Saga too was determinedly chewing the tough meat. When Diya's gut-brain let her go, and she regained some agency, she tapped Saga's elbow and nodded toward the huddle of COP refugees. They brought the second half of their meals over, pulled up stools, and listened.

"I was just saying," a stocky white man with a Californian accent said, giving Diya and Saga a nod, as if catching them up on his train of thought was doing them a real favor, "that given the rumblings from the Equatorial Common Interest, we may want to prepare now to have a proper geoengineering studies track next year."

"A couple heat-stricken countries sending up a dozen planes full of sulfate does not 'geoengineering' make," replied a Japanese-Australian woman Diya knew in passing as Mirai Hase, an expert on ocean acidification. She looked sour not only about the damp but about getting cooped up with the likes of the Californian. "Real solar radiation management requires constant global maintenance. No one expects the ECI to last any longer than New Warsaw, the European Reconfederation, Latinamérica Unida, the Commonwealth Climate Convention, Pax Solaris, or any of the other half-baked internationalist schemes people keep trying. If they actually did anything, it would be like slapping your television to get an instant of clear picture. Unilateral geoengineering is just a movie cliché."

"Okay, sure," the Californian allowed, "but whatever you call it, there's a solid chance they'll dump *some* particulates, just like China tried. We shouldn't be caught flat-footed a second time. If there's a jolt of the system, we should try to capture good data on it, even if it doesn't give us more than a weird chilly year and a few crop failures. How effective were the dispersion strategies? How quickly were changes felt? How rough was the snapback after? The people trying to do SRM for real know an ECI run would be just a trial balloon—excuse the pun—but they need to learn from it if they are going to make geoengineering a priority after Reunification."

"Reunification is fantasy," a gray-haired man, already deep in his cups, put in. He had a heavy Muscovite accent. "At best incoherent conspiracy theory. At worst problematic political talk that will make many countries crack down on climate dissidents—including perhaps us in this room, da? You Americans should know this better than anyone, Noah."

"Granted." The Californian, Noah, impatiently tried to steer the conversation back on track. "But just as a thought exercise, we should ask what it would look like if we human beings made *good* choices. Maybe we can't fix anything, but if we *were* going to fix it, what would need to go right? Well, Reunification would probably have to really take off. Or at least some kind of defragmentation, whatever your politics. Then what? Then we implement real, effective SRM to buy us some breathing room while we set up the demilitarization-to-drawdown pipeline. Okay, say that happens. What are our new failure modes if we get that far? Well, one, we fuck up the SRM. Maybe it doesn't work well enough, or it backfires. Either way we waste a bunch of energy and political capital dealing with that instead of building the carbon removal apparatus. So, what can *we* do? We can study the ECI attempts, if they happen, and generate some lessons learned."

"*We* aren't all going to study the same thing, no matter how pressing," Mirai said. "Scholars do research, submit papers, we organize the good ones into tracks. We can curate them a bit, but really, we aren't on the granting end of this. We don't direct research. This kind of naiveté about how actual institutions function is exactly why no one takes the Reunification crowd seriously."

"But what's the point of all this, then?" Noah waved his arms in exasperation. "Why still have a COP if we aren't going to play to our outs? Maybe we can't fix the planet, but why do climate research at all if we aren't going to try?"

"Because people need knowledge to *survive*," Mirai said, which got some approving nods from the group. "Billions of people need to know where it's safest to live, what crops to plant, what water sources and ecosystems they can depend on. Those choices are informed by the data *we* collect, share, and peer review at this conference. If the planet breaks, and we can't fix it, we don't get to buy a new one or trade it in at the shop. We have to keep living on it. And *we* have to do the research that will help us know how. If you aren't going to contribute to that, you shouldn't be here."

This was all pretty standard COP political talk. In a way Diya found it a comforting return to normalcy.

Mirai was right, of course, but Diya thought she was being perhaps a bit hard on the Californian. She liked the look of him and she liked his hopelessly unselfconscious demeanor, even if he was intellectually out of his depth in this crowd. Throughout the conversation, Noah kept glancing their way, checking for approval. But then Saga got up to return her plate to the bar, and Diya saw his gaze follow the Swede's tall, blonde form. Ah well.

4PM

The bartender kept turning the volume up on the ancient speakers, to keep the thrash-cumbia playlist audible over the howling wind. Eventually the COP crowd complained that they couldn't hear themselves talk, and he just turned the music off entirely. Everyone shut up for a minute and listened. They heard the high-pitched whine of arrhythmic gales, the hum and rattle of glasses vibrating in the sink, the occasional bang as loose objects smacked into the building.

Diya went to the bathroom again. In the hallway was a window. The bars and bullet shutters on the outside had prevented flying debris from shattering the glass. She unlatched the window, and it swung open violently, forcing her to jump back. With a little handle, she cranked open the rusted slats, squinting against the wet air blowing in.

She saw first the torn-up train tracks, covered with rushing water. The flood was not so high, but there would be no leaving until it receded. Not that she'd want to. Across the tracks a craggy tree had been ripped out of the ground and speared through the windows of an apartment building. Old power lines had fallen and were sparking in the water; Bastardo's batteries must have kicked in—a rare case of something going right. She tilted the shutters higher. Rain stung her face. The sky was still seething.

Back in the bar, Noah had cornered Saga with a trio of new beers. He handed a glass to Diya as she sat down.

"So, what are you working on?" he asked them both. It was a common scholarly greeting. Unlike "Who's your advisor?" or "Where do you teach?" it didn't assume any particular academic status. It also avoided revealing that the greeter didn't know who you were, even when they probably should.

"Cross-checking data and records," Saga said. "And, ah, picking brains, when necessary."

"When necessary?" Noah put his hands to his skull in mock terror.

"My collective is attempting to gather a full archival history of the climate collapse. But of course, with fragmentation, it is very difficult. The RINGO group does a good job of distributing transcripts across their network, but everyone is bringing limited and contradictory findings. We would get the papers, but not be able to reconcile them. So, we decided to send an archivist here to the COP to collect unpublished raw data and ask clarifying questions."

"Saga is being modest," Diya put in. "There are so many charlatans in our field now. They take advantage of the gaps and decay in our peer review process. We are very lucky Svalbard is so aggressive in rooting out bad data that they sent her to us. I have seen her get a fast-talking geophysicist to admit he flubbed half his numbers with just the force of her glare!"

"Wow, a dashing, climate-archival investigator!" Noah enthused. "What do you do if they don't give you their data? I'm imagining you grabbing their papers and sprinting away from a herd of angry RINGOs like Indiana Jones. Do you steal ice cores from government vaults as well?"

"Really, most of what I do is ethnography," Saga said. "We can't contextualize what RINGO sends us without understanding the culture of the COP."

"What about you?" Noah turned to Diya. "What are you working on?"

Diya pointed up at the rattling ceiling. "This."

Noah leaned forward. "A meteorologist? At the COP? I never thought I'd see the like!"

"Someone has to do something useful with all the mushy data these people gather," Diya laughed.

"Diya's models predicted the storm," Saga said. "That's why we ended up here. We left the venue before the soldiers arrived."

"Wow, nice of you to tell the rest of us!" Noah said, but Diya could tell he was still joking, flirting.

"Were you there after the drones?" Diya asked. "I assumed it would make securidad lock everyone down, just in case the attack was an indicator of our importance. 'Half of value in conflict is posturing' and all that. How'd you get away?"

"Bribed some guards," Noah shrugged. "They let us slip out in the chaos. Getting a drink seemed more pleasant than wherever they were taking us. So how bad is the storm?"

"On the Katrina scale, maybe a point four. I don't have proper instrumentation to say for certain. It's notable for where it's hitting, not how hard. I'm more worried about what it stirs up than what it knocks down. Hot planet, dirty peace, you know."

"People always say that," Noah said. "But there's every indication that South America is stabilizing, just like Africa, and the Oceanic conflicts. Despite the population boom, demographics are rounding the corner. Lots of places don't have enough young men left to keep fighting. Material constraints are normalizing, so there may actually be *less* appetite for scrapping over resources. I mean, who wants to die for oil anymore? Water? Even places like this bar have circular filtration now—they have to! Land? Holding land just makes you vulnerable. Timber? We already cut all the forests down. Minerals? There's hardly the industrial apparatus or the international markets to make mining worthwhile."

"What a pollyanna you are," Saga said. Diya could tell she was annoyed but too polite to show it. "I suppose we're just a few degrees warming away from utopia, at this rate."

"It's not utopian to say the trends point to an upturn in the cycle of history. You think the world was any less messed up at the end of World War II? And then what happened? Much of the globe experienced decades of relative prosperity, freedom, and cultural renaissance. People were sick of violence, so the wars stayed cold. The UN was born—the First Unification, we'll call it, because it can happen again!"

"Or," Diya pulled out a cig and lit it, "the second half of the twentieth century was a historical anomaly, an outlier just as unlikely to be repeated as this storm. Prosperity was powered by cheap fossil energy. Freedom was a correction, the result of oversized empires downsizing. Cultural renaissance depended on a growing media landscape that's now stagnant. The only thing I can say for your theory is that it's too vague to be peer reviewed."

"What can I say? I'm a futurist. Vagary is part of the futurist toolbox—it's the only way to keep looking out further and further."

"A futurist?" Saga said, sounding interested in spite of herself. "One of those visited our collective once. The historians hated him, but the rest of us found his ideas quite provocative."

Noah gave Saga a winning smile and a wink. "That's us! We're a rare breed these days, I admit. Only people who believe we have a future bother hiring us. But the future is still coming and, as Ms. Hase said, we still have to keep living in it. Just because we might not conquer the galaxy or achieve full luxury communism in our lifetimes, doesn't make thinking about the future any less interesting!"

"You know what I think about the future?" Diya said. Something the Californian had said had suddenly churned a bit of bile up into her mouth. She felt like spitting. "I think the future is mostly going to be like the past. Most human beings throughout history have lived scared of the sky and scared of each other, and they still do. 'Nasty, brutish, and short,' you name it. Peace, plenty, consensus reality—none of these are guaranteed. There's no cycle, just a few decades' exception to that general rule. An exception that now we are paying for dearly, as history seems determined to make us remember our place."

Diya got up and stalked back to the bathroom. She immediately felt a bit bad about snapping at Noah. Bad weather and confined spaces had a way of fraying her temper. She decided to make an effort to knit her nerves back together. She lit another cigarette and stared out the window, watching the storm pound the world outside.

7PM

By evening everyone in the bar had gotten bored, so dinner became a group affair. The networks were all down, and the tattered books and magazines lurking in the corner booths were all in Spanish. The batteries had run down. For a while a gas generator had kept the lights on, rumbling from the floor below, but when it became clear the city's grid would be out for quite some time, the staff dimmed the lights to save on fuel. So for dinner they pushed together a few square tables and lit their meal with little candles and a clatter of oil lamps.

The other patrons had slipped off, perhaps hanging out on Bastardo's other floors, eating with the staff in the kitchen, helping with tasks to ensure the building's community survived the storm. The COP crowd, being foreigners who largely didn't speak Spanish, had stayed huddled in the bar, alienation sapping their agency. Diya suspected that the Argentines in Bastardo knew what kind of military moves might get made after the storm but felt that explaining it to the foreigners was not worth the hassle. A generation earlier the language gap might have been less pronounced, but multilingualism was always discouraged in periods of jingoistic fragmentation. So the bar staff treated them a bit like children, encouraging them to stay in the main room, to stay a little drunk. Eventually food was brought out and set up for them family style.

"A toast to COP60, or however you count it," Noah said, raising his beer. The mood, though brittle, had taken a convivial turn. They toasted.

"I never did hear the full story about the years that got canceled," Saga said.

Mirai ticked off on her fingers. "Let's see, I know we don't count COP25 as missed, even though they had to move it at the last minute. COP26 was postponed a year by the first Covid pandemic, but they kept the number when it finally happened—squeezed in before the Scexit chaos in Scotland. There were a lot of those change-ups. The first one we actually skipped numerically was COP31, right?"

"Yes, it was supposed to be in Bangkok," said Shi Ann, a Canadian sociologist Diya had met a few times, who up until then had been quiet. "But the Thai monarchy blew itself up and the junta canceled all the flights at the last minute. Really set things back, and the next couple years were very demoralized. I remember the Glasgow Pact fell apart for good at COP33, so they skipped COP34 and 35. It took two years to bring the parties back to the table—most of them, anyway."

"So why'd they keep counting those years? Instead of postponing, like they did with Covid?" Saga asked. Diya suspected Saga knew the answer, but posing these sorts of questions was part of her ethnography: checking for consistency in shared understanding of the institution's history.

"I think everyone wanted to maintain a sense of things moving forward," Shi Ann said. "Even if they weren't. A lot of publics then were already unclear what the point was. It would have been a bad look to get stuck trying to organize the same event year after year."

"Did not want numbers to become cursed," the Muscovite, Borya, added. Diya had chatted with him a bit and learned that he was a geochemist who'd been coming to the COP almost as long as she had, though somehow they'd never met.

"Sure," Shi Ann allowed. "Screws up the branding, the diplomatic significance attached to specific anniversaries. All that stuff was pretty important during that era."

"The other reason," Diya put in, "was that RINGO and some of the other constituencies managed to have meetings those years. So, there was some continuity. In fact, I think you could argue that those years laid the groundwork for the nonstate continuation we are a part of today."

"Well, a toast to that!" Noah said, apparently annoyed that everyone's attention had drifted away from him. They toasted.

"Then there was a pretty good stretch, if I remember right," Mirai said. "Going right up to the UN dissolution in 2039. So then there was no COP44, I think."

"COP44 was skipped," Diya said. "But not entirely because of the dissolution. The community had seen that coming—everyone did—so they made preparations for a meeting. A Conference of the Constituencies, though no one wanted to call it 'the COC.' But then a rumor spread that the constituencies meeting was part of a shadow-UN conspiracy. Glo-nats threatened to bomb the meeting. The organizers canceled, and the next couple years they held the meeting much more under the radar with a different name. But everyone who showed up still called it 'the COP,' so eventually we got back on the original numbering system."

"This is why no one wants to hear your Reunification talk, Noah," Borya said, not unkindly this time. The history lesson was softening everyone up. "Global Nationalist Front would make trouble again, if they thought COP was Reunificationist plot."

"Makes sense," Noah said. "After all, if a real, viable movement to overcome the incoherence of glo-nat fragmentation were to emerge, it would probably have to start in one of the few remaining institutions that brought together smart people from around the world to examine the state of our planetary system."

He looked around the table pointedly. Everyone busied themselves with their meals.

"What about the last two missed COPs?" Saga prompted after a few moments of silence.

"Let's see," Shi Ann said. "COP50, Los Angeles. UCLA got occupied for a few months by that cult, the Burning Tower or something. The organizers didn't realize how dangerous the cultists were until right before the COP. Had to scramble and shut it down last minute. Then, a few years ago, COP56 coincided with the Cape Town water riots. We had exactly one session before the windows started getting smashed in by bricks. I remember because it was *my* session, and all the way up to the roof I kept thinking to myself, 'at least I got to give my paper on the marginalization of refugee communities in import-substitution agricultural transitions.'"

"Honestly, missing only two out of the last fifteen isn't bad, even though now we are only a conference of a few hundred, instead of tens of thousands," Mirai said. "Interesting that it's mostly been these black swan–type disruptions that actually prevent us from meeting. They make up only a small portion of the world's dangers, but the more organized the violence, the easier it is to plan for it and pick a different city."

"I suppose the same is true of climate shocks," Saga said, pondering. "RINGO avoids putting the COP in places vulnerable to wildfires or hurricane seasons. A neverstorm like this is the exception to the rule. Come to think of it, are we the first time the COP has been hit by a climate disruption?"

"All disruptions are climate disruptions," Borya said. "Thai junta, climate disruption. Glasgow failure, climate disruption. UN dissolution, climate disruption. Glo-nat bombs, climate disruption. Cult, climate disruption. Water riots, climate disruption."

"Not sure all of those claims would pass peer review, but it's a good point," Shi Ann said. "To Saga's question, I'm not sure. We might be the first time the weather itself shut down the COP."

"I had an uncle who did climate scenarios modeling," Noah said. "He told me once about going to an academic conference in the teens that got snowed out by a freak blizzard. Everyone just slogged back to the hotel and watched TV. The experience didn't change anyone's scenarios."

"I think there were a few times early on when climate disasters hit somewhere in the world at the same time the COP was happening," Shi Ann said. "It was still unusual enough that there was a lot of media attention and activist speculation that seeing the crisis unfold live would galvanize the negotiations. Sadly, it didn't."

"Of course not!" Borya scoffed. "Experience is not ideology. Is not economic or political incentive. Seeing the problem does not ensure a will to act."

"I was a listless uni student, waiting to inherit my family's money, when I saw Cyclone Vayu destroy Mumbai," Diya objected. "I'd say that experience was crucial to giving me an interest in weather and climate. It steered me away from business and partying. It's what brought me here today."

"Experiencing the damage done by Thor's Year, particularly the loss of some of the great Nordic libraries, is part of why I joined Svalbard," Saga agreed.

"Outliers, outliers," Borya said dismissively. "Noah has had California sunshine every day of life, and still he is here. Millions hurt by climate disaster are not."

"Hey, I grew up breathing wildfire smoke just like everyone else," Noah said, raising his hands. "The question isn't just what you experience. It's what you *choose* to do with it. *We* can choose what to do with this experience too. We can let it motivate us to new action, or not."

"You think we will be different?" Borya laughed and took a pull of his beer. "We will drink together at our hurricane party. Then, if we are lucky, we will get to go about our business. Nothing will change. Same as every COP for last sixty years. Such is our shame."

"That's not fair, though," Shi Ann said. "It's apples and oranges, a totally different moral landscape. We're just researchers. The negotiators who failed to achieve a stable climate treaty should surely be held to a different standard. They understood the urgency *and* there was potentially still time to act."

"We don't know either of those things for certain," Diya said. "But it hardly matters now. Excusing their failures doesn't get us anywhere."

There was a general round of head shaking and grim agreement. Lamenting the sins of their predecessors was a time-honored COP tradition. Diya gnawed at some of the fried meat. She wondered if it was from some cow grown on razed Amazon ranch land. She wasn't supposed to eat beef, but she'd long since stopped being picky.

9PM

The wind still howled, but they began to sense that the storm had rounded a corner. The rattling of the building was not quite so violent. Or perhaps they'd simply gotten used to it. Diya did not know. On a smoke break she peered out the slatted window into wet and fearsome darkness. Briefly she fantasized that the world outside had always been this way, would always be this way. She felt as though she had lived through this storm over and over again. In a way this was true: what was to separate this storm from the one that had destroyed Mumbai three decades earlier, or any of the others she had chased in the years since? The same molecules cycled through every gust and cloud, getting angrier and angrier as the sun poured its energy into their hot house, afflicting every nation, faction, race and creed in one way or another. One big, endless, shared storm.

When she finished her cigarette, the others had cleared away the plates and poured themselves new drinks from behind the bar. After so many hours, they had begun to feel like they should have the run of the place. Then, as if by unspoken agreement, they all drifted back to the table and picked up the conversation more or less where they'd left off.

"So where did it all go wrong?" Saga asked the group. "Or, to put it another way, what needed to go right?"

"Are we assuming, for the sake of argument, that climate collapse wasn't locked in before the COP began?" Mirai said.

Saga nodded. "Yes. Maybe some of the crisis was unavoidable, but surely we can all agree there was a window where the right actions could have made some kind of difference. To borrow Noah's phrase from earlier, what would it have looked like if we had made good choices?"

"Al Gore wins the Florida recount in 2000," Noah said immediately. "He comes to the COP with a budget surplus and a public eager to unite after an exhausting election, and he actually pushes for *more* ambition, not less. He gets the US into the Kyoto Protocol. If 9/11 still happens, he uses it to snub Saudi Arabia and push a clean energy bill to get America off foreign oil. Renewables tech gets fast-tracked and spread worldwide, cowing the fossil fuel industry. Emissions start to seriously drop by the end of his first term."

"Whew, yes, let's get the Americocentric fantasies out of the way quickly, shall we?" Mirai said. "If only Hillary had won, we would have stayed on track. If only Biden would have got a second term, we'd have successfully course-corrected. If only Yang hadn't been assassinated. If only, if only. The United States is always just one good president away from leading the world into a shining utopia. Never mind that if Gore had wanted to push harder, he had eight years to do it with Clinton. Never mind that people thought Obama would save the planet, and then the US blew up years of work at Copenhagen."

"She's right," Shi Ann agreed. "I don't think rewriting political history however we want is that useful of a thought exercise. You might as well say, 'What if the Soviet Union never fell, and instead Gorbachev read Murray Bookchin and led the world into a harmonious ecosocialist utopia?'"

Everyone looked at Borya to see if he'd register an opinion, but he just shrugged.

"Maybe it's more in the spirit of Saga's question to ask what *we* could have done better," Diya said, feeling a bit sorry for Noah and wanting to redirect the discussion. "*We* meaning the COP, the negotiators and organizers, the researchers and activists, everyone who preceded us in this institution."

"The parties could have allowed the IPCC to do a real study on the dangers of 1.5C warming much earlier," Mirai said. "It seems quaint now, given what we're locked in for, but the 1.5 report really did change the conversation for a while—there just wasn't enough time to act on those new targets. But if we'd had the report a decade or so earlier . . ."

"I don't know. Wasn't the shocking timeline exactly what briefly activated people?" Shi Ann said. "Anyway, the fact that there was so much resistance by some of the big countries to even *talking* about 1.5 seems to me to point to a deeper problem with the framework."

"Montreal Protocol," Borya said, rapping the table with his knuckles. "It was blueprint for completely different problem. Built in many terrible things—demand for consensus, no enforcement mechanism."

"I think you're onto something," Saga said. "Fixing the ozone layer was possibly the peak internationalist achievement, yes? A clear agreement at Montreal and a smooth transition. Maybe the framers of the UNFCCC got overconfident. I don't blame them for wanting to build on what worked, though."

"But everything was different about climate change!" Mirai said. "The politics were different. The economic impacts were different. The freaking *chemistry* was different. It was crazy to think they could just use the same process."

Borya nodded vigorously.

"And they had a clear symbol to help people understand the stakes," Shi Ann said. "The picture of the hole in the ozone layer. It was new and scary. Climate change was scary, but storms and droughts weren't actually new. And our crowd kept changing the name of the problem—global warming, climate change, climate crisis, everything change, everything crisis. Nothing stuck."

"It always comes back to communication," Mirai nodded. "The research community thought they could just do the science, and people, governments, someone would act. Maybe they believed staying above the politics would preserve their credibility. But what they needed to do was wade in and learn how to talk to the public about what the science *meant* for their lives."

"I disagree," Diya said. The conversation was starting to swirl around the same tired tropes of ivory tower boffins trying to explain the onrushing peril to ignorant, oblivious chuds. "The decades of obsession with 'climate communication' were born out of a lack of analysis of actual power. Montreal worked because DuPont Chemical's patent on Freon was about to expire. They went along with the transition—profited from it. The fossil fuel industry saw decarbonization as an existential threat and immediately started confusing the issue, politicizing the science, resisting the environmental movement on all fronts. It was a bigger sector, more powerful. Capital as a whole was deeply invested in fossil profits and reserves, and in the end sided with the industry at every turn. Not having a 'picture of climate change' to show the public only mattered because the technocrats weren't willing to go against their monied masters."

"A lot of good that did them," Noah said. "I don't dispute that the fossil companies encouraged early fragmentation, even directly backed the glo-nats and some fash. But their business has to be a lot harder now that you can't build a pipeline across borders without someone trying to blow it up. Or ship your product across the ocean without an escort against pirates. Shit, neo-Peronist-style governments have renationalized a nice chunk of carbon reserves. Fragmentation isn't working out for anyone, even the bad guys. Everyone gets that, and if they don't yet, they'll figure it out quick once a real discussion around Reunification starts spreading."

This brought a few eye rolls, but it was Borya who said what they were all thinking.

"And yet, military tanks and humvees still get their diesel. Rockets and helicopters do not run out of fuel. Fossil companies are still in business when many rivals are not. Gears of war turn and will keep turning until no one is left alive to fight."

"A global stocktake of current carbon extraction activity is basically impossible," Mirai added. "The best we can do right now is extrapolate from atmospheric and oceanic numbers. For all we know, emissions today could be accelerating faster than at any other point in history."

There was a pause as everyone sipped their drinks and contemplated this bleak scenario. For a moment Diya wasn't sure if the rattle and banging in the distance was debris snapping in the storm—or gunfire and mortars, half-covered by the hiss of rain.

"Whether fragmentation was deliberate or not, getting the whole world to agree on a single protocol and vision was probably impossible," Shi Ann said. "Climate impacts are too heterogeneous and unequal, and the types of transitions required were too diverse. What if the environmental movement had focused on regional treaties? Or sanctions on polluters? Or built climate action into smaller trade deals? Those kinds of wins could have cascaded. We needed a domino theory, instead of trying to build a house of cards."

"I like a good mixed metaphor as much as the next fellow," Noah objected. "But just because one version of a global approach failed doesn't mean nothing could have worked. Instead of nation-states as the negotiating parties, we could have pushed to organize the COP by bioregions or industrial conglomerates or network tribes or markets or anything. Westphalian sovereignty was on the decline at the turn of the century, but the COP propped up an archaic system that couldn't solve planetary problems."

"Or that might have just created different axes of fragmentation," Saga said. "Before dinner you spoke of historical cycles. If Reunification is inevitable, maybe fragmentation was as well, and every approach was doomed to failure. Or, at least, every approach that relied on cooperation across borders and solidarity between ethnic and religious groups."

"Non-ecofascist approaches, you mean," Shi Ann said, before Noah could get a word in. "Those fuckers like cycles too. 'Strong men make good times. Good times make weak men. Weak men make hard times. Hard times make strong men.' Except whatever year it is, *they're* always the strong men, and everyone else is the disposable weak."

Diya swirled her drink, remembering the feeling of déjà vu that had gripped her at the window. She shoved her glass to the middle of the table.

"Seems to me this conversation is going in cycles," Diya said. She began to dig in her bag for her smokes.

"Am I the only one who feels like they've heard all this before? Earlier tonight, or yesterday, or last year, or at every COP for sixty years."

"I feel that too," Saga said, cocking her head. She gave Diya an odd look, then glanced at Noah. "I've seen it in the archives. Decades of the same discourse, running in circles, looping back in on itself. We never come to consensus or real conclusions. Everyone just talks about what people *should be* talking about, while outside the damage piles up."

The discussion hit a lull. Diya finally fished out her cigarettes and tucked one between her lips. She tossed the pack on the table and waved to the others to help themselves. Everyone took one, though only Borya and Shi Ann lit up with her.

"You know," Noah said, glancing at Diya. He examined the cancer stick in his fingers like he'd never seen one before. "I've been thinking about what you said earlier." Then he added for the rest of the table's benefit: "Diya really tore me a new one for fetishizing 'the future' as a stand-in for progress. I guess I wanted to say, I call myself a futurist, but I know people have been thinking about 'the future' for ages without it ever arriving. History happens. Social change happens. Climate change happens. 'The future' doesn't happen."

Diya nodded. "Where we are today must have seemed so far away to the first RINGOs at the first COP. I don't think they even imagined there would be a COP60. Or even a COP43. They must have thought, 'Well, we'll work things out or we won't, but one way or another this shouldn't take that long.' And of course, things didn't work out—whatever that means—but I rather suspect that even if they had, some of us might still be here arguing over some of the same questions. Or at least arguing over something. Just not quite so drunk, only."

This earned a rueful laugh from the table, and they toasted again with the dregs of their drinks.

Midnight

When the storm quieted, the fighting began. They heard first the rumble of diesel humvees, sloshing through the waterlogged, debris-strewn streets. Then shouting, machine gun rattling, the bang of a grenade or an IED. They heard wails of pain and terror, followed by long minutes of quiet, then more wails, more terror.

The maneuvers had the plodding, organic rhythm of human warfare; it was still raining too hard for most drones to be effective. It was easy to imagine the opportunity the gangs, milicias, and insurgent groups saw in the storm: power out, networks down, drones incapacitated, securidad spread thin. Maybe they were moving on long-desired targets—fuel or water storehouses, government offices, or buildings of cultural significance whose capture or vandalism would earn clout for a particular political narrative. A block away on Av. del Libertador were museums which memorialized the people kidnapped, tortured and exterminated by the US-backed junta during Argentina's first Dirty War. Or maybe they were just lashing out, looting, acting on vendettas that had been sworn that very night.

The party atmosphere was gone in the bar. Everyone tried to sober up, kept their bags at hand. They took turns at the window, looking down on the train tracks. There was not much to see. The storm surge had mostly receded, and the flood trickled around the tracks ankle deep. The window looked toward the Universidad de Buenos Aires and the Río de la Plata. They had no angle on the rest of the city. From the occasional boom that rumbled the glassware, Diya imagined a skyline blistering with little explosions.

Noah and Borya crept downstairs to confab with Bastardo's staff. They returned to report that the building was empty. Either the Argentines had slipped out or they had bunkered in some basement the foreigners couldn't access. Pounding on the locked doors on the lower floors prompted no answer, and Bastardo had survived enough conflict to have very strong doors indeed.

This prompted a discussion about what the group should do. There were half-hearted suggestions of making for the campus, for the airport, for the military base twenty kilometers to the west. No one had particularly good reasons for any of these destinations, but talking about doing something foolish felt better in this anxious hour than doing nothing at all.

In the end they decided to stay put. They barricaded the door to the bar. Hours passed. They tried to sleep, but no one did.

Then there was a hot streaking whine and the biggest explosion they'd heard all night, rocking the building's foundations. They rushed to the window. This time there was something to see.

Rockets were arcing into the city from out over the river. Brazilian rockets from occupied Uruguay, they had to assume. They slammed into the waterfront, the campus, the museums. Little waves of heat shook the rain.

Now hiding out in the tallest building for several blocks seemed deeply unwise. They argued for several more minutes. More rockets came, closer this time. They shouldered their bags and fled inland.

Without power, with clouds covering the moon and stars, the city was just lines of shadow in wet darkness. Sometimes truck brights or muzzle flashes bounced off the buildings and the water, shocking their night vision. They splashed as quietly as they could, keeping at the edge of the street. They paused at intersections to check for fighting, which seemed all around and yet not quite anywhere at the same time. Diya clutched her cigarettes in her hand. Behind them rockets rocked the coast.

Diya's déjà vu was gone. Instead she felt a thudding, hammering, pounding sense of not knowing what would happen from one moment to the next. The idea of a future where this could all be fixed, or a past where it could have been avoided, or another history where good choices had been made—all that seemed hopelessly naive in the chaotic, gray night.

Ahead they saw trucks parked across the road, figures stacking chunks of debris into a more defensible barricade. They zigged a block, then zagged another. They heard the occasional pop of sniper fire, but couldn't quite be sure of its direction. The fighting was clearly moving away from the rockets as well. But what else could they do? They kept going, lifting their hands to show they were unarmed whenever they passed men in fatigues or men with guns or silhouettes turning to them in the dark.

The rain was slowing. Diya prayed it wouldn't stop. Clear skies could mean drones.

Once Diya tripped over a body. She shined up her phone to make sure it wasn't, somehow, one of her own group. The man—boy, really—had a wide face and wore bloody fatigues. It was, she realized, the kid who had sold her smokes from the kiosk near campus, L. Soto. He was slumped against the wall, breathing heavily, hands applying limp pressure to a gunshot wound in his gut. She knelt to examine him. He looked at her, and she thought she saw a little recognition in his eyes.

Diya wanted to do something for him, but she felt Saga tug at her arm. She pressed the half-empty pack of cigarettes into his lap. Then she moved on. There was nothing to be done.

Shared Socioeconomic Pathway 1
The Green Road—Sustainability

IF WE CAN DO THIS,
WE CAN DO ASTEROIDS!

Second Sunday

After the storm came a bright, blue morning and much work to be done. Faces poked out of broken windows, waved at each other. Glass was swept off doorsteps. Leaves were cleared from solar panels, and extension cords were tossed from rooftops to charge phones and run water filters. Gardens were picked through for surviving plants and produce. Dusty disaster kits were pulled from closets. Rooms were found for those put out. Neighbors mobilized cleanup crews. Strangers went to check on strangers.

The city's disaster management office unfurled a network of command nodes and block captains to coordinate recovery efforts, verify and spread information, and triage requests for help or resources. As good as this operation was, it—like all centrally planned human activity—did not quite map onto the actual social relations emerging on the ground. So adjustments had to be made: cleanup crews merged or split, given impromptu legitimacy, new nodes established and a few disbanded, natural leaders brought into the process, captains asked to step back when they fell short. This was normal and natural but occupied a surprising amount of time and energy compared to the actual physical labor of cleanup and recovery.

On the whole, however, the awkward melding of top-down planning and bottom-up improvisation had distinct advantages when done right. Each paradigm saw and corrected the blind spots in the other's strategy. The street knew the newcomers, the live-in boyfriends, the visiting relatives, the not-quite-common-law squatters. The state had yearly census records that included the antisocial, the shut-ins, the infirm, the workers of grave-yard shifts rarely seen by neighbors. So in this way the recovery efforts reached and checked in on most everyone who'd been living in Buenos Aires during the storm.

There were, in the final count, very few injuries and only one death: an expecting mother who died of complications during childbirth when she went into labor during the storm and was unable to reach either hospital or midwife. The baby, who survived, was delivered by strangers at the cafeteria where the woman took shelter. The woman was mourned throughout the city, memorialized in murals, her name given to a program of free lay-midwifery trainings offered to the public months later.

Despite the unexpected ferocity of the storm—which coalesced out of an unstable pressure system and arrived via an erratic and swift path that left little time to prepare—the city's infrastructure held up well. The grid had been fortified by Transition Era retrofits that spread solar generation and energy storage throughout the city. These had been won by robustness hawks who had shouted down pinchfist technocrats at raucous public planning debates during the height of Mobilization Politics. The grid was clumped into "cells" that could function like independent microgrids if connections were torn. They could also deliver power to neighboring cells if generation was uneven, as when rooftop solar was damaged or obscured by debris. The balance of the grid algorithms meant that there was quite a bit of incentive for friendly competition between cells to keep maintenance up and energy demands down. The hyperlocal expertise that approach encouraged came in handy after the storm, and by the end of the first day, grid-nerds had jerry-rigged fixes for almost every house and business that did lose power. And where the grid held, electricity still flowed from the city's pump-storage vaults, which not only helped contain the deluge but netted power from captured stormwater.

The limited flood damage was another win that the robustness hawks would be crooning about for years to come. They had built the "slanting garden" embankment parks high enough to hold off a storm surge from the Río de la Plata, as unlikely as that had once seemed—as

well as a significant amount of sea-level rise, should the
projects to stabilize and rebuild the planet's glaciers fail.
The city's once-notorious drainage problems had been
much improved by shifts to porous roads and a general
greening of public and private urban spaces. Rooftops
channeled the downpour into rain barrels. Flowerbeds,
trees, and mossy lawns that had been parched by the
summer heat soon bloomed a deep green.

Perhaps the most notable forms of damage were
broken windows—which spilled music onto the streets
for several days after—and breached ground fridges.
Many home-clusters and apartment buildings had invested in these passively cooled cellar pods, subsidized
by the city in the late '20s. The storm coincided with
the sealants on doors and joints in the most popular
models showing their age, and some porteños found
after the storm that rain and groundwater had leaked
in, generally making a mess. While little food had actually been spoiled, a good cleaning and repair was
usually in order.

Many moved the contents of their fridges out to the
sidewalks, laid out on collapsible tables, and encouraged neighbors, cleanup crews, and passersby to help
relieve them of their perishables. Others brought out
produce salvaged from wind-torn gardens and window
boxes. Much of this bounty was used to prepare huge
communal meals to support the recovery efforts. The
days after the storm became a kind of informal feast
week—the smells of cooking overpowering the whiff
of flood rot, the noise of construction and demolition
mixing with sounds of cooking, laughing, sharing food,

dancing, belching, singing, dishes clinking and bottles clanging as boisterous toasts were made.

The joy and catharsis of these festivities, however, was cut through with a serious mood the storm had brought to the city. While everyone agreed that the damage could have been much worse, the black swan disaster was a reminder that—despite the successes of the Transition Era, the great reforms made locally and globally—they were not out of the woods yet. The Earth's oceans and atmosphere were still dangerously energized by greenhouse warming and would remain volatile for generations. Glaciers were melting, tundra was thawing, ecosystems were struggling to adapt. The climate crisis had an inertia that would take centuries to turn back, even if the most critical transitions had been accomplished. The planet was in for a long, rough, precarious time. Everywhere in the world knew it, but some weeks some places knew it more than others.

This collective feeling was heightened by the fact that the global planetary management negotiations happened to be taking place in Buenos Aires when that angry cloud-bank had charged up the Río de la Plata. Still called the "Conference of the Parties" despite the shifting nature of the UN's framework convention, COP60 became a topic of much local interest in the week after the storm. The public-facing parts of the conference were mobbed. Live videos of negotiations and roundtables were shown at bars between fútbol matches. COP attendees—both porteños and visitors—were feted at community clean-up dinners, invited into classrooms, approached on the trolley, bombarded with attention, questions, opinions, requests, demands, accolades, anger, congratulations, business proposals, and romantic propositions. It made for a very lively conference.

The increased public engagement brought more energy than usual to the more provocative elements of the otherwise anodyne negotiations. This included thorny questions about the risks and benefits of solar geoengineering, about the costs of various computationally intensive atmospheric modeling moonshots, and about the optimum temperature to stabilize at when the world's carbon disposal industry eventually brought greenhouse gases under control. This last—where to set the metaphorical global thermostat—was of particular faddish interest. "Cuál es tu número?" folks would ask, and be met with shouts of "doscientos setenta!" and "trescientos doce!" "Muy frío! Trescientos cuarenta y cinco, por favor!"

Badge holders would get asked to weigh in on this debate when spotted at salons, breakfast cafés, church services, or parks. And because opining is a form of thinking, the half-dormant working groups that handled these questions suddenly found themselves enriched with new ideas.

All this activity—the cleanup, the grid repairs, the feasts, the window and fridge replacements, the midwifery classes, the interest in the COP, the engagement with the big questions of planetary management—none of it was smooth or evenly distributed. The effects of the storm played out over months and years, different from neighborhood to neighborhood, sprawling beyond the city, the continent, the single conference. There was no single story of the storm any more than there had been a single cloud or drop of rain or gust of wind. The storm wove into countless stories, playing a role in triumphs and disappointments, breakups and flirtations, marriages, divorces, promotions, bankruptcies, career changes, affairs, ambitions, artistic breakthroughs, friendships sparked and ended, children conceived and come of age, homes remodeled, blocks transformed, cities reinvented, life carrying on as it always had into a future that was forever uncertain but not, it was hoped, unwelcome.

Second Monday

Noah hung from a trolley strap, hands blister-sore from the cleanup. The sensation put him in mind of his childhood—growing up rough around hard work he'd since gone too soft for. Born in the late teens, his parents were among the first to join the American ecosystem repair efforts, even before the jobs program had ramped up. So he'd been a Green New Deal baby through and through: rambunctious, curious, mirroring the energy he saw all around him, raised by the whole Climate Crisis Corps village in the Montessori classrooms of the great, fire-prone Californian forests. The grown-ups would chop and haul brush for the Big Cali Clear, building a sparer landscape that could sink carbon without burning out of control. Noah would sit watching, demanding to help until someone gave him a pile of sticks to move.

He'd drag the tinder-dry saplings to the sledges, rawing his fingers, scratching his legs and arms in the untrailed woods. He remembered feeling so proud but realizing, later, that the task had been more about keeping him safely out of the way of more dangerous work. There had been little pressing need to move those sticks.

Thus it had been the day after the storm. Like many COP attendees, Noah had offered his help to the clean-up crews assembling outside the apartment building where he'd been given lodging. They had appreciated his eagerness, but he and the other COP-goers—being foreign political operatives—had little of the hyperlocal knowledge or technical skills necessary to be truly useful. So they had found a lone pile of rubble and debris for Noah to help move—an old church that had somehow not been brought up to code, whose north wall had collapsed when a tree toppled. There was no urgency to this job, but afterward the crew captain thanked him and inquired concernedly about keeping his strength up for the negotiations. Noah got the hint. He decided to stay out of the way and get back to the job he'd traveled a hemisphere to do.

So Noah got to the negotiating room early on Monday. The building was a union hall—Noah's home turf. He didn't know enough about the Latin American labor scene to parse the acronyms on the door, but from the look of the muraled walls Noah suspected the venue belonged to the building trades.

Most of the COP sessions were spread throughout Buenos Aires—though never more than a couple of trolley stops apart. This kept the proceedings accessible and helped keep the negotiators from getting insular. Now, however, some of the spaces were being double-booked to help organize the block-by-block recovery efforts.

Noah shouldered past tables where men and women in coveralls were handing out mold masks and protective gloves.

Saga Lindgren, of course, was already in the session room, sitting at the table, shuffling through notes, wooden name block already stood on end to request the floor as soon as they started.

"Good to see you, Noah. How are your children?" Saga said when Noah plopped down next to her.

"Sad to learn the storm didn't wash me out to sea," Noah said. "How's the art?"

"We will see on Wednesday, I suppose," Saga said. "Are you coming to my show?"

"Are you going to help me get a straight answer on 2100 stabilization targets? You know the taigal parties have been holding out on me."

"And *you* know I'm here negotiating for the arts constituency, not the Nordic block."

"Doesn't mean you don't have influence," Noah pressed. "I bet if you only let people willing to name a target into your show, we could whip a couple votes."

Saga gave him a reluctant smile. "Flattery, flattery. Personally, I think they aren't unreasonable in wanting to wait. We don't actually know what a stabilized 350 ppm world looks like. We barely have meaningful data on the preindustrial climate. You are asking people to commit to setting the thermostat while the climate still has a lot of chaotic inertia to spin out."

"I don't care about the thermostat," Noah said. "I care about planning the work that needs doing to get wherever the world wants to go. Everyone says, 'Well, if we don't like the climate, we can always draw down more, right?' Wrong. The final target determines the number and size of the reservoirs we eventually have to build, and the pipelines to get there. Which in turn determines the resources we allocate to building them, how many workers we hire, how many pensions we budget for. And it can be dangerous work, so there's health costs to figure in as well. And that's just my concerns representing the carbon trades unions. The other trades will need to know the scale of the operation, so they can plan to get us the materials and equipment we'll need—and everything else. Dragging our feet on these decisions loses us a lot of efficiencies."

"That's the problem with a planned economy, Noah," Saga said, clearly getting into the spirit of the debate. "Plans change. The big data projects are still crunching the habitability numbers. Maybe lowering the temperature loses us more land near the poles than it gains us around the equator. Maybe we'll be able to reclaim land from sea-level rise, maybe not. We just don't know. If we make big bets now and have to change course later when we know more, the costs might be much more than what we pay by asking for a little flexibility from the unions."

"So set an ambitious target now and we can pull back a bit later," Noah said. "A vote not to make a clear plan is a vote to leave it up to some kind of nebulous market, and we all know how *that* pans out for workers. Last thing we want is a boom-and-bust cycle when the stability of the planet is at stake."

The other negotiators and observers started to file into the room.

"The delegates you are asking me to convince represent people whose farms may eventually freeze over if the COP votes for targets under 300," Saga said. "I think we should take their concerns seriously."

"If they want me to take their concerns seriously, they can always join my union. We've got plenty of jobs that need doing." Noah shrugged. "Anyway, what about those guys? Some of them are getting 55C summers, massive desertification. A little ice age might make their lands arable again."

He waved at the group of African equatorial negotiators taking their seats. They looked intense, dashing, confident—as always. With a small surge of envy, Noah remembered the previous year's COP in Nairobi. The city was verdant and shining, the result of bleeding-edge leapfrog development. It was one of the great ironies of the Transition Era that, when funding started to flow, many areas in the global south had been able to adopt sustainable economies and technology much more quickly and readily than the richer, earlier-industrialized countries

While Lagos and Kinshasa transformed into "green megas," and new arcology cities sprang up in Africa's solar surplus zones, America and Europe still slogged through endless rounds of retrofits, sprawl repair, densification, redevelopment, leakage analysis, further retrofits. Noah's "developed world" found itself weighed down by a crumbling layer of materiality that had once been a great achievement but, it turned out, hadn't been built right the first time around. California was up to its knees in undead housing and zombie infrastructure: clunky, toxic, half-functional buildings and systems that polluted the landscape and got in the way, but still usually had to be lived in. Finding the talent to drag them out of the "wreckage of the unsustainable" was a mounting challenge; America's best architects and planners were brain-draining themselves to Africa and Asia, where more ambitious and glamorous projects were possible and plentiful.

These different paths to sustainability played out at the COP as differences in aesthetics and attitudes. Noah often found his mild, childish jealousy mirrored by looks of pity from the equatorial delegates.

"I thought you didn't care where we set the thermostat," Saga said, nudging him.

"I don't," Noah said. "But the carbon trades membership voted to make bringing home clear 2100 disposal targets one of my negotiating priorities, so here I am, trying to push the conversation along any way I can."

"Okay, okay," Saga gave him a friendly, placating pat on the shoulder. "I can't make promises today, but I do know much of the Nordic delegation will be at the reception before my show on Wednesday. If you are looking for a place to work your magic, I can get you on the list. *If* you bring your union friends to my show."

"See, there we go," Noah said, tapping his name block to hers like a comical high five. "I scratch your back. You scratch mine. The system works."

Second Tuesday

The fair—and thus the world—seemed to Luis to be full of wonders. He walked among the pavilions, scanning the wares and sampling the foreign foods on offer. Around him buzzed COP attendees and porteños, drifting toward smells or demonstrations, getting pitched on this or that UN program or national effort. The party pavilions had taken advantage of the storm to reconfigure themselves early for the end-of-COP fair. It was one of the great boons to the city hosting the conference: diplomats, scholars, and activists from around the planet brought with them casks of goods that were usually hard to get outside their bioregion.

The negotiations traditionally concluded with a sharing of this "global bounty" brought in special by COP delegations. This year the storm—which Luis had watched blow through from inside the Villa 31 community center where he had been reporting back to his constituents on week one, and which now seemed like a memory from an ancient past—had prompted the organizers to move up their timeline. They hoped consumables from the fair could make their way into the post-storm feasts popping up around the city. Luis had bowed out of his negotiation commitments for the morning to help out at the Buenos Aires city booth. Of course there was very little foot traffic at the BA booth since all their contributions to the fair could be found at every grocery and restaurant in the city. So Luis had in turn taken leave to wander through the fair, which had always been his plan.

He sampled fruit from Southeast Asia, beer from Western Europe, grain cakes from North America. He fingered exotic fabric weaves and toyed with artisanal electronics. He watched religious paraphernalia he didn't recognize being passed out to grateful believers.

At the Indian pavilion an older woman in a sari tapped out small shakes of spices onto the back of his hand. He licked them off, one by one.

"I know this one," he said in English, puzzled. "Or something similar. I think I tasted it before, on a trip to São Paulo. Or maybe Paramaribo?"

"Very possible," the woman said. "Much of northern South America is in one of our sister biomes. Star anise, cardamon, turmeric—all spices we've worked to global-ize into the rest of the pluviseasonal tropics. Is this your first COP fair?"

"My first COP, actually," Luis admitted. "Luis Soto, Argentine youth delegation."

"Diya." She gestured namaste. "I used to be a RIN-GO, then I was with the Indian delegation, then the UN. Thank you for taking time to be here at the COP and this fair. I know when a neverstorm struck my city, when I was about your age, nothing seemed to matter besides being out on the streets, doing what I could."

"I'm definitely . . . distracted," Luis admitted. "The community I serve took the brunt of the storm. But my constituents are very keen to understand what's happen-ing at the COP and wanted me here, instead."

"Good! Very sensible of them. How are you finding the fair?"

Luis, not wanting to come off as wonderstruck as he'd been feeling, cast about for something insightful to say.

"Well, it has me wondering how tricky it would be to get my hands on some of this stuff were the COP not in town."

"Far from impossible, I should think," Diya said. "There are a few of the old, refrigerated container ships still running on diesel dispensations, and more and bigger wooden sail-ships and solar drift-ships launch every year. You might have to put in a special order or sit on a waitlist. At worst get mildly lucky in a lottery. But there's very little, within reason, that you couldn't get your hands on with a little effort."

"I suppose," Luis mused. "My parents sometimes complain that, back in their day, they had access to anything and everything from around the world, within days and with just a couple clicks—not that they could necessarily afford it. But then they also tell me I'm spoiled by how much better the food is now."

Diya laughed. "My family could always afford whatever we wanted, and still half of what we ate was too processed, dried, and stale to be enjoyable. All so we could have the supposed luxury of eating blueberries in Mumbai all year round. How boring! Much better to always have the novelty and nostalgia of the next season's dishes around the corner to look forward to."

"Of course! My parents also complain about old clothes and electronics. Everything wore out or broke down after a couple years, unless you could splurge on stuff that wasn't made in Chinese injection-mold factories or Bangladeshi sweatshops."

"Everything you find here will be durable," Diya assured him. She had a professorial air about her, which Luis, who had been thinking of going back for another round of schooling, found congenial. "When you aren't trying to flood every market on the planet, you can afford to use artisan methods that expend more energy and time on smaller batches. The math works out because you get products that last longer and have less distance to travel."

She offered him a wooden tray with sliced guava, as well as pieces of flatbread, some already spread with a yellow chutney, others with something red and spicy-looking. He sampled the food and found it as good as everything else he'd tried.

"So how is everything at this fair so fresh if it came from across the planet?"

"Oh, that's not very difficult—just energy-intensive. Too much so to necessarily do at a huge scale, but we make an exception for the COP. Plus everything comes from the finest crops grown in each bioregion. All to grease the wheels of diplomacy, right?"

"I suppose I owe you a vote or two for all this, then," Luis said, taking another couple of flatbreads from the tray.

"I'll let it slide." Diya winked at him. "These days my role is mostly ceremonial. I do this and that, but I'm involved in only a few actual votes."

"Like what? I'm still wrapping my head around everything that goes on here."

In the eight days that Luis had been part of the Argentine delegation—liaising between the negotiations and the former slum community where he organized the annual needs-census—his biggest takeaway was how sprawling planetary management was. He'd met people who worked on glacier engineering, soil carbon overclocking, rural depopulation facilitation, megafauna outreach and negotiation, keyline design, desertification rollback, sea retreat community planning, carbon mangrove reclamation, rare greenhouse gas accounting, indigeneity onboarding, ecosystem services cryptocurrency stabilization, orbital commons stewardship, the search for sustainable extraterrestrial intelligences. He only knew what about half of those things meant, and even then only in the vaguest terms, but he was determined to finish the COP with at least a sense of scope.

"Well, today I helped organize this fair around the sister biomes 'seed and skills share' programme," Diya said. "We help communities adopt a kind of 'globalized bioregionalism.' This means diversifying production and agricultural practices with crops and techniques native to other parts of the planet that share similar climatic and landscape characteristics. If we're successful, you may see more exotic foods at groceries, coming in via overland trade from the other biomes on the continent."

"What about invasive species?"

"That's a paradigm we're moving away from. We've already globalized the planet—there's no going back. The question is whether we can globalize the useful and the beautiful, not just the weeds and the pests. There's no rolling back the extinctions that have already happened, either. What we can do is weave together many different lifeways to make our ecosystems truly healthy, rich, and robust. Biodiversity, not nativism. So that means a lot of careful swapping of bugs and birds and molds across continents, in addition to seeds and skills. It's painstaking work, but I like to think the institutions doing it are finally starting to knit together something like the post-nation-state political imaginary. Eventually we hope to have enough solidarity and coherence within and between bioregions to drop the concept of 'countries' for something a little less prone to conflict and contradiction. But, of course, these big goals have to start somewhere, and I think they start with food. Ah!"

A pair of blue badges arrived and showed Diya a tablet.

"It was very nice to meet you, Luis," Diya said, offering him another friendly wink. "Give my friends here your contact. Perhaps we can continue the conversation later in the week. Enjoy the rest of the fair."

Luis did so, and then she was gone, whisked out of the Indian pavilion and into the crowd. Luis stood there, looking after her, mulling over what she had told him and wondering if and how he could find a place in such grand ambitions. Eventually one of the other pavilion minders came by and offered him a cup of chai. Luis accepted, and sipping the sweet tea roused him from his reverie.

As he walked away, Luis felt a hand close around his elbow. It was Cheeto, a wiry Belgian youth delegate Luis had befriended in passing the first week of the COP.

"Man, did you just get fifteen minutes of one-on-one conversation with Madam Kapoor?" Cheeto demanded. "You gotta teach me your secret."

"What? What do you mean?" Luis asked. "You know her?"

"Brother, Diya Kapoor is the former executive secretary of the UNFCCC. Which means she basically used to run the planet!"

Luis started at that, but it made some things click into place. He felt an urge to chase after the older woman and thank her properly, call her "Madam Secretary," maybe apologize for wasting her time. But there was more fair

to explore, and his own booth to check on, and then back
to negotiations in the afternoon. He would simply have
to go about his day, holding on to the notions she'd left
in him: that new imaginaries were possible, that small
things could be part of big plans, and that the powerful
and accomplished were not so different, up close, from
the likes of him.

Second Wednesday

As the minutes ticked by before the gallery opened, Saga
fiddled with the lighting. It was, like so many things,
an ever-evolving process. She had gotten it perfect two
weeks earlier, but then other works had been moved into
the space, changing the cast of the walls and the shape of
the air. And so she'd had to tweak the lighting algorithms
over and over again until the color and intensity of the
illumination hitting her piece—the centerpiece—was
back to how she wanted it. By the evening of her show,
she worried that the many adjustments had muddied
her own sense of her original vision.

Nonetheless, when the gallery doors opened and the
early arrivals trickled in, she set aside such anxieties,
pocketed her control gloves, and prepared herself to present her art—however imperfect, but good enough. She
smiled and mingled, made small talk about the conference as needed, though mostly she tried to keep herself a
bit apart from the rest of the COP, a bit aloof, and instead
directed the guests' attention to the various contributions
of the arts constituency arranged around the space.

Saga liked quite a few of these pieces. There was the
longitudinal series of oil-on-canvas sunsets, painted by

a bot every day for twenty years and curated to highlight the slow changes in the sky that came with falling emissions—and the chaotic weather that still persisted. There was the slab of bleached coral reef, delicately extracted from the ocean floor and hung on the wall without context. There were the provocative videos depicting a utopian society living in a hothouse climate, children joyously exploring decimated cities, lovers embracing in the light of whale oil lamps, families picnicking to watch tornados. And her favorite of the pieces not her own: an exquisite iron sculpture of a runaway train spilling over a cliff, caught and righted by many hands and broken bodies, carrying on its way.

The reception got more crowded, buzzier. In the mix Saga saw friends and acquaintances, a few other artists. Noah Campbell was there lobbying members of the Nordic bloc. More came in, a wave of dignitaries Saga was slightly shocked to see attending. She paid respects to indigenous leaders and the archbishop-elect of Buenos Aires. At the last minute, Diya Kapoor swept in with a young Argentine in tow, who was trying very hard to keep the deer-in-the-headlights look off his face. Saga managed just a brief, "So lovely to see you, Madam Secretary," and then heard in her ear the cue from her assistant that it was time to begin.

Saga excused herself, changed in the back, tugged on control gloves. When she returned to the gallery, the mood had mostly hushed.

"Comrades, we have been lying to ourselves," Saga called to the crowd, the only frame she would give them for what came next.

She stepped to a platform next to a wide, roped-off basin in the center of the gallery, then began the sequence. Fog started to pour out of the ceiling above the platform. Working her gloves, she formed it, from several meters back, into a grey, rotating sphere. The tiny jets in the platform and ceiling, combined with a nearly undetectable projection layer, caused the ball to begin to resemble a monochrome miniature of the Earth. Then thinner vapor circled the globe, forming several sets of cloudy shells, each highlighted by projector light in a different color: a translucent planetary nesting doll.

It was an impressive visual effect. The piece used a great deal of computing, along with projections that flickered so fast they confused the eye in very specific ways. The eyes could even focus *through* the globe, onto each of the different layers, much the way the ear could pick out one voice in a choir. It took some getting used to, and as the crowd's eyes adjusted, they oohed and ahhed.

Saga didn't need to control the installation live. Indeed, for the last several days of the COP, and for a month after, while she lingered in Buenos Aires, the piece would run automatically for visitors to the gallery. But there were advantages to being there, doing some of the shaping and controlling the timing. It allowed her to instill a sense of theatricality. For instance: Diya Kapoor's young companion was by the ropes, and when he reached out a finger to touch the fog globe, Saga triggered the next stage.

His hand pulled back, startled, rebuked, impressed. Images had begun to resolve on each of the layers, one by one. The onlookers stirred with appreciation. Each layer showed video of a different part of the living world: trees swaying, grain waving, lava sliding, the lights of skyscrapers coming on and off, insects roiling in a carcass, soil shifting with time-lapsed microbial action, fur rippling over animal muscle, a thunder-cracked storm. And more. All wrapped around each other, forming one harmonious system.

This, Saga knew, would have been perfectly acceptable to many galleries. Some in the crowd were no doubt thinking she was making a statement about the interconnectedness of all beings. With her opening pronouncement, some would interpret the piece, thus far, as a reminder that humans were just one part of a complex web of life, a truth forgotten beneath civilizational lies. Something sophomoric like that.

But then Saga triggered the third stage. The images panned or shifted, and became violent. Sometimes in abstract ways—a fiery clash of colors and shapes. Others were less subtle: rioters fighting with old-school police, pirates firing on an oil tanker, gas canisters pinging off laser-lit streets, a tractor pulling down a statue, children kicked by jackboots, tasers jolting into picket lines, strikers beating scabs.

The crowd pulled back, stunned by the abrupt change in tone. Saga let the horror show play for just a few seconds, then flipped the images back. The onlookers again leaned in to see the fine details of the many-shelled world, though more wary this time. She twisted her hand—more violence. Then back—green hills and rain in puddles, famous arcologies rising out of familiar skylines. Then violence once more.

She went through this cycle a few more times, and then she made each layer of vapor puff away, until all that was left was the dense Earth of fog, on which was playing an incoherent mash of all the other layers—peaceful and not. She let this roll as she made her way off the platform toward the gallery's back exit. Eyes followed her, but most stayed watching the globe. She snapped her fingers and the cloud dispersed, the projector lights went out.

Saga had a glass of water, changed back into her reception dress, and then went back out to see the already dwindling crowd.

There was a small smattering of applause when people noticed her return. She made her rounds through the room, thanking the clumps of guests for coming. The gallery emptied. Eventually only one group remained, Noah Campbell and a few others Saga knew. They were discussing their interpretations of her performance, which Saga would prefer not to engage, but Diya Kapoor was there, and she felt obliged to tell the famous climate leader goodnight.

"For me it evoked a very retro feeling," Tara McVey, another American, was saying. "Like it was referencing all those expectations of apocalypse and collapse. There really was a fear, before the Transition Era, that everything would fall apart really quickly and violently. People had trouble imagining how smooth the transition could be, how much sense most changes would make when governments were stirred to act. I think a few of the shock-and-awe images were from old movies."

"For me, I think it reminded me of the opposite," Noah countered. "We have this sense now of, like, 'of course

things worked out this way.' But really, they didn't have
to. We could have made worse choices, or lost a bunch
of coin flips no one remembers anymore. And I think we
have this gradualist narrative. That things were smooth,
like you said. But that's because we've already erased
from our collective memory a lot of the worst of the rup-
tures. It took real struggle to finally get climate action.
Direct action, strikes, even a little violence."

"Well, of course," Tara said. "But compared to the
horrors of the twentieth century, the climate movement
was extremely nonviolent. There were sit-ins and oc-
cupations and general strikes, but no assassinations or
bombings, at least not in the US. Just peaceful protests!"

"Yeah, but my parents were at all that stuff, and they
always say there was no such thing as a nonviolent protest—just ones where the protesters didn't fight back."
Noah shrugged. "Plus, nobody likes to talk about the
pipelines that got sabotaged, or the threats and harassment that got coal plant operators to desert. But making
carbon energy companies harder to operate was a big
part of why investors eventually threw the old fossil
giants under the bus, trying to buy themselves time. And
reimagining politics produced a lot of hate and distrust. I
remember growing up, we'd occasionally encounter sad,
sick people who were camped out in the woods, sure the
government was about to come euthanize them as part
of some sinister population control program. Then when
the population started to dip a lot faster than people had
expected, there were even more conspiracy theories."

"So that's what you got from the piece?" Tara said,
nonplussed. "That the Transition Era was messy?"

"What I got was a reminder not to feel so smug. We talk like it was all part of some inevitable enlightenment. But The Enlightenment was a culture war too, with its own share of violence. So, to me, that's what the 'lying to ourselves' bit meant. And maybe a warning that we aren't immune to rupture now. If we forget the past, that kind of antisolidarity could come back again."

"I don't think anyone my age forgets the strife and tension before—and during—the Transition Era," Diya Kapoor said. "Luis, you're the youngest here. What is your view?"

"Madam Secretary," said the young Argentine, hair in a prim bun. He seemed to be choosing his words carefully. "I'm no expert. But my impression is that history has been a long process of stepping back from the edge. Once, many wanted to use nuclear weapons, and they had to be talked down. The time before the Transition Era was another edge. So, I'm thinking about the people I organize with, who I know so well. And I ask, do they have it in them to walk up to the edge, whatever edge it may be? Yes, I think they do. I don't think that goes away."

"Well put," Saga said, stepping a little closer to the group. A couple of them started.

"Since you were listening, can you tell us who got it right? About your intent?" Noah put in.

"That's a terrible question to ask an artist," Saga said. Then added, "But I will say none of the videos are from old movies."

"Regardless," Diya said, in a tone that made it clear she was ending the discussion. "Saga, my dear, that was wonderful."

"Thank you, Madam Secretary."

"Just one question," Diya added. "Does the work have a title?"

"I don't like titles," Saga said. "But in the promotional materials the gallery has circulated, it's called 'Nuestra Tormenta Compartida.'"

Noah started to fumble with his translator, but she stopped him.

"It means, 'Our Shared Storm.'"

Second Thursday

Diya hated talking to rich people, but she was good at it. She was one herself, or had been, though that sense of isolated entitlement never quite leaves you, she feared. The lingering rich needed most to be made to feel that they were winning, in charge, going of their own free will, even as the sea overtook them. So that's what Diya offered them.

"This, my esteemed friends, is the kind of glory your money can buy."

Diya stood at the prow, shouting to be heard over the wind and the waves and the low hum of the sail yacht's electric control motor. Her audience sat on cushioned benches bolted to the deck of the boat. They drank mimosas and wore gold "VIP" badges that glinted in the summer sun, an ego-stroking touch Diya was particularly fond of.

She waved at the octagonal structure looming ahead of them. It looked impressively industrial, in that very twentieth-century way. But it was also draped with greenery, vertical crops hanging in sheets from four of the sides. Around the structure the open ocean was broken by smaller works—a farming flotilla of rafts and buoys, beneath which hung yet more crops: kelp, scallops, mussels, fish traps, and soil bags growing a dozen kinds of artisanal aquatic vegetables. It was one of the more impressive offshore agriculture projects in the region, providing significant fish protein to nearby Buenos Aires and helping reduce local acidification levels in the surrounding waters. But Diya wanted to keep her audience's attention on the rig.

"The platform you see before you began life at a shipyard in Itaguaí, Brazil, at the cusp of the Transition Era," Diya continued. "It was destined to be an offshore oil drilling rig pulling toxic hydrocarbons out of the Argentine Basin, at the behest of a hungry market and hungrier investors. But we have found a better use for it. Mr. Campbell?"

Her audience turned to Noah, who grabbed hold of a rope and hauled himself up to stand unsteadily beside her. She had brought Noah along to explain the technical details of the storage project, but also to remind her guests of the powerful unions they might come up against if they said no. She would be the carrot, Noah would play the stick.

"Far below us, under the ocean floor, is a large, porous formation of sedimentary rock," Noah explained. "Right now those pores are filled with saline—salt water. With robots and special concrete-setting microbes, we have fashioned that formation into one of the world's first carbon waste reservoirs. Carbon dioxide is transported here in a flexible undersea pipeline from an air capture plant tethered to the offshore wind and solar farm a few dozen klicks further out. Here the CO2 is pumped down into the reservoir, forcing out the saline into the ocean and pretty much staying put. The technical details are obviously more complicated, but I promise you the chemistry is too boring to be worth getting into. The gist of it is, we take clean energy, use it to fix waste carbon out of the atmosphere, then put that sky trash more or less back where it came from—underground, where it contributes to neither radiative forcing nor ocean acid-ification. Questions?"

"Why do all this, instead of planting more trees?" asked a man with thick plastic sunglasses—showy and expensive, given the limits on nonessential plastic man-ufacturing.

"As I understand it," Noah said, "that's an ongoing de-bate at the COP—the balance of these strategies, anyway. But one answer is nutrient bottlenecks. We've got a lot of waste carbon, but that's not true of everything we'd need to do huge amounts of afforestation. Another is land, which people don't always want to give up to plant carbon dark forests. Plus, because of the sensitivity of weather systems, if you plant a new forest in one spot, it can reduce sequestration in a neighboring area. A third answer is time. Industrial air capture works somewhat faster than trees mature.

"And finally, when trees eventually die, they release much of the carbon they captured back into the air—usu-ally on a shorter timeframe than we are looking for with carbon storage. That's fine when you're working at scale.

You count the forest, not the trees, as it were. Still, forests catch fire, trees burn, and then you're set way back on your drawdown. Living systems take a very different kind of management. Nothing wrong with that, but we think it's better to put as big a chunk of the problem as we can away for good, and not all in the tree planting basket."

"Why the pipeline?" someone else called out. "Why not just do the capture right here?"

"Eventually, yes, we hope to incorporate generation, capture, and disposal all into the same facilities. But right now these pieces are largely being built out in a modular way while the carbon trades find their feet. The other reason is that we might want to pipe CO_2 in from other sites, depending on the eventual capacity of the reservoir and where the solar surplus shakes out."

"You don't *know* the capacity of the formation?" A bottle blonde in the back raised a skeptical eyebrow. She wore a high-fashion version of the jumpsuits coming out of the new European clothing provision houses—a statement of either envy or scorn for the empowered masses, Diya didn't know which.

"It's hard to know anything for sure about anything that far underground," Noah said, unfazed. "This isn't some big cave we've dug. We're talking about rocks, under more rocks, under the ocean. But we have sensors, we know where the carbon goes and whether it stays there. The biggest challenge now is building an organization that can ensure the integrity of those sensors and the data coming from them, and be financially responsible for any carbon leaks that occur over the minimum time we want the carbon to stay put. Say about five hundred years. Which, I guess, is where you all come in."

Diya took the prow again.

"Esteemed friends, you know I have brought you here today to show you the vital work funded by the Planetary Trust. This is but one of hundreds of beautiful, state-of-the-art storage sites we are building. They are true marvels, a great gift to all the world and every living thing in it, and to a hundred generations yet to be born. We are also funding a great deal of the previously mentioned afforestation and countless other projects that benefit the planet as a whole. But when something benefits me, I pay for it. When something benefits a city or a nation, that city or nation pays for it. Who pays for something that benefits everyone? We need a new kind of institution, one whose mandate is both broad and long. That is why most of the parties to the UNFCCC individually—soon to be followed by the UN as a whole—have instituted a global wealth tax that pays into the Planetary Trust."

The mention of taxes made the crowd shift uncomfortably.

"I know, I know," Diya said, giving them an understanding smile. "A topic sure to ruin an otherwise lovely day out on the yacht. That's why I'm here to offer you an alternative. All of you control significant private assets, and while your investments have been smart, much needed, even world changing, we now have ever more data showing that private mobilizations of capital are deeply inefficient for achieving long-term climate stability.

"We need to put the world's capital into the hands of the Planetary Trust if we are going to build projects like the platform you see before you and operate them for the next five hundred or one thousand years. And we need that money fast, because, esteemed friends—we are still up against it. The storm our fine host city experienced this week is a reminder of the tipped-over world we are desperately trying to right. Every year that passes with this much carbon in the air continues our planet's slide toward the hothouse. We need every resource available to us to build the removal industry at scale and at speed!"

At this Diya stepped down from her perch and took up a champagne flute of mimosa. She held it up, as if making a toast.

"My most esteemed friends, today I ask you to make this possible. Hand over your assets to the Planetary Trust, so that we might accelerate our plans and stabilize the world. Why wait for the wealth tax to siphon them away year by year? I know, as well as any of you, the burden of these vast, clunky masses of capital. Masses that many of us never asked to be charged with keeping. They are in their own ways as toxic as the oil this rig had once been built to extract. Relieve yourselves of them, put them to better use. And in return, you will be cared for all your life, with freedom to go and live as you please, a citizen of every country party to the Trust. You will be honored forever on these monuments for your generosity. You can build us a stable climate future. And if we can do this, we can do asteroids! We can handle the many dangers that lurk in deep time. The Planetary Trust can ensure a prosperous human future where your names will be remembered!"

She swept back up to the prow and pointed at one of the massive struts lifting the platform above the water, which had just come into view. On it were freshly carved names—famous names of ultrarich people Diya had already talked out of their fortunes. Diya raised a toast once more.

"To you! May your names be honored for a hundred generations!"

She drank. Many of the others drank with her. Those who did not glanced away, not able to meet her eye. She'd get them too, soon enough.

Diya's speech was done. She did not mention how paltry the perks and pensions and honors were compared with the titanic sums they'd be giving over to voluntary democratization. She did not mention the increasing legal precedent for holding the megarich accountable for what their investment portfolios paid for in terms of fossil extraction, deforestation, ecosystem damage, and political dithering. The Hague's climate trials had a momentum all their own now, with prosecutors always hungry for new enemies to feed into the environmental justice maw. She did not mention what she would hint at later, in private conversations: that the best way to avoid a dangerous audit was to just give up their money now, after which prosecutors would look the other way. She did not mention that the unions Noah was representing were clamoring for the Trust to move forward with more hostile expropriations of such "stuck capital."

Noah caught up with her on the ride back.

"Heckuva pitch," he said. "If I were a lonely, anxious billionaire, I'd be jumping to give you my money. Though it leaves a sour taste in my mouth, seeing their egos stroked like this. They are my class enemies, after all."

"There's only brief catharsis in seeing your enemies humiliated," Diya said. "Letting your enemies save face, however, can prevent them from becoming your enemies again. Noah, understand, these people used to basically run the world. Now we are, shall we say, laying them off from that position. Today's theatrics are just the difference between saying 'you're fired' and saying 'we're letting you go.' If that difference helps them slouch quietly into the night, I say we let them have their dignity."

"Still, it rankles. Why should some rich assholes get their names on that strut, instead of the workers who actually built the thing?"

"Because the world isn't fair, Noah. Not just yet, anyway."

Second Friday

The closing ceremonies had birds. They swooped through the open air, in no particular pattern, black dots sharp against the cloudy gray sky. Some landed. Great cawing corvids gathered in a shuffling murder on a hill to the side of the lawn in the park by the river where foreign badge holders and many porteños had come to put COP60 to rest.

They were a new addition to the proceedings, these birds. A week before, biodiversity sensors had noticed that the city's avian population had largely fled the path

of the neverstorm several hours before humans understood its trajectory enough to begin battening down the
hatches. So a pilot project had been launched to explore
incorporating local birds into an early warning system
for future neverstorms. Experts on nonhuman relations
who happened to be attending the COP were quite en-
thusiastic about this project, and they had used their
translation algorithms to liaise with local crows and
ravens. In the end the conference organizers, as a favor
to the host city, had invited the corvids to attend the
closing ceremonies as a show of good faith.

There hadn't always been closing ceremonies, just
closing plenaries where those who came sat next to
their luggage, sending messages, winding down the
hours until they needed to head to the airport. But the
introduction of voting had brought a raucous, contentious dynamic to the climate negotiations, an energy
to match the politics of the spiky, fraught, ultimately
successful Transition Era. So the closing ceremonies—
as well as second Thursday parties—were added as a
way to smooth over the little grudges and frustrations
inherent to any democratic process and remind everyone that they were all, ultimately, striving toward the
same goals.

Speeches were made, sermons were given. Ceremonial drinks and snacks were handed out and consumed in
unison, songs were sung—awkwardly but with genuine
feeling. The birds even got in on the act, circling up, it
seemed, for their own set of pompous speakers offering
long-winded remarks. It was all a little silly, but it had
the intended effect of helping everyone leave more or

less on good terms, investing them in the process for the year to come.

Noah, Saga, Diya, and Luis were all in the crowd, feeling varying degrees of annoyance and bemusement at the proceedings. They did not stand together, but with others in their delegations, or other friends and acquaintances they knew from their years at the conference. A couple of times one would catch the other's eye, perhaps nod or smile. When the ceremonies were done, Luis found Diya and thanked her again for spending so much time with him. Diya thanked Noah for his help on the yacht. Noah jostled into Saga as they made their way out, cracked a joke. Saga told him to bring his kids next time. Luis watched Saga's yellow hair drift off toward the exit, and he resolved to take his sister to see Saga's installation at the gallery. None of them were so important to any of the others. But there was a congeniality there, a solidarity, a sense of being together as small parts of a larger whole.

The COP the following year was in Vienna. The year after that, Lhasa. Then on to Moscow, Cairo, Seattle-Vancouver. The process continued. The last of the world's greenhouse emissions were chased down and either shut off or accounted for. The carbon disposal industry ramped up its operations. Glaciers were kept from sliding into the sea and then, decades later, were rebuilt. Bioregions leached away national sovereignty. The wealth tax slowly leveled the world's inequalities, funding the Planetary Trust to first manage the climate, then draw down ocean acidification, then scan the heavens for dangerous objects, then provision basic needs for each person born on the Earth. All these were efforts that had their own momentum, but the COP remained a meeting for various political players to come together for two weeks each year to iron out the kinks in these projects, resolve disputes, and decide collectively what kind of world they wanted to live in.

None of which was to say that things were perfect, or that history had ended, or that the planet turned placidly until the end of time. There were still setbacks. There were still terrible climate horrors waiting in the second half of the century: millions displaced, crops lost, species driven to extinction, the world's resources and planetary boundaries more than once strained to the brink. There were still culture wars and a few shooting wars. There were still diseases and preventable deaths. Many of the realities that had made life so hard for so many for as long as humans had walked the Earth still persisted. To be born was to know suffering—that much never changed.

But there was a sense, even as all these tragedies were lived through and dealt with in turn, that human life had passed through a bottleneck, or perhaps the eye of a needle. Some would always ache for the urgency and purpose of the Transition Era, and the dangerous, uncertain time before. But most recognized that, on the other side of this great test, a wide space of possibility had opened up before us. Life did not have to be lived in the shadow of onrushing doom, or with a sense of guilt at the damage one did by simply existing, or consumed by anger at the sins of a greedy, foolish past. There were so many ways to live, so many scenarios of human being to yet explore.

On that last day of COP60, in the hot December of 2054, there was only a hint of such possibility in the air. But the hint was there, as surely as the birds, and the river, and the lively clouds rolling in from over the water, promising a gentler rain.

AFTERWORD

Speculative Fiction, Climate Fiction, and Post-Normal Fiction

The stories you've just read, which constitute the lion's share of this book's intellectual effort, are works of fiction. Specifically, works of speculative fiction. All fiction, of course, speculates and imagines. Even the most realist of realist writers are still putting on the page events that did not happen, creating people who do not exist, hypothesizing about the feelings and ideas those made-up events prompt in the minds of those made-up characters. What distinguishes speculative fiction is the systematic exploration of not just imaginary people and their individual stories, but imaginary places and societies, imaginary eras and historical events, imaginary biology and physics, imaginary creatures and species, imaginary gods, imaginary religions, imaginary politics, imaginary technology, imaginary climates, imaginary ways of life. One need not create all aspects of a story from whole cloth for it to count as speculative fiction; rather one must merely set one's story in a context recognizably and deliberately imaginary.

Speculative fiction often gets categorized as belonging in one of two camps: science fiction ("sci-fi") or fantasy. Science fiction tends to speculate about a future that might happen—worlds that could one day be. Fantasy is usually, though hardly exclusively, set in a past that we know didn't happen—all the worlds that could never be. Fantasy worlds are often made fantastic by the existence of physics-breaking forces called "magic." Science fiction worlds, on the other hand, are usually made fantastic (when they are fantastic) by advanced technologies that are not yet available but nonetheless seem plausible. Depending on how plausible those made-up technologies seem or are made to seem, science fiction may get called either "hard" or "soft," and the aesthetic character of those technologies and the worlds they create leads to a proliferation of other subgenres, from cyberpunk to space opera.

Of course, all of these categories are somewhat contested, both from within the genre fiction world and from without. I use "speculative fiction" as an umbrella category for both sci-fi and fantasy—as well as horror, alternate history, "weird fiction," and other genres that lean heavily on the imaginary—and to make room for differences in approach between different kinds of fiction that think about the future, as I will discuss shortly. However, "speculative fiction" has also been used (rather infamously, if you are a sci-fi lover) by writers such as Margaret Atwood who wish to distinguish their imagination-driven work from the lower-brow tropes and stylings associated with mainstream science fiction. "Speculative fiction encompasses that which we could actually do. Sci-fi is that which we're probably not going to see," Atwood said in a 2009 interview with *Wired*. While Atwood's writing and thinking about the environment and climate have certainly influenced my own, I must clarify that my use of "speculative fiction" does not intend to make such an arguably pejorative distinction.

Genre bickering aside, the five stories (or novelettes) in this volume are pretty much science fiction of the "hard" variety. They take place in the future (a bit more than thirty years hence), and they contain technological and cultural apparatus that make some sense given our current technological and cultural trajectories—although each story imagines those trajectories going in somewhat different directions. There is no magic in these pieces, aside from the vague sense some characters have that the crucial moments of their stories are significant or familiar; this is more me exercising my poetic license than a statement about how space-time actually works.

But perhaps a more precise way to categorize these stories is to call them "climate fiction." Climate fiction (often called "cli-fi") is fiction that centers the Earth's climate and in particular the changes coming to our climate as a result of anthropogenic forces.

What are these forces? To summarize: since the beginning of industrialization human beings have, at an ever-accelerating pace, dumped massive quantities of greenhouse gases into the atmosphere—particularly carbon dioxide, the low-energy waste product of burning hydrocarbon fossil fuels like coal, oil, and natural gas. More greenhouse gases in the air mean that more heat is absorbed by the Earth and less radiates out into space, increasing the "radiative forcing" that determines whether the sun's energy makes the Earth get warmer or cooler on average over time. Warmer average global temperatures, even rising by what seems like a small amount, produce big and varied changes on our climate and ecosystems: hotter summers, longer droughts, more powerful storms, more erratic weather patterns, the melting of polar ice leading to the rising of the seas.

Why should fiction concern itself with this complicated thermodynamic process? Because, to quote Atwood (2015), it's not just climate change—"it's everything change." Climate change affects where we can live, what we can grow. Its effects destabilize the socio-technical-ecological systems we depend on to provide our basic needs and comforts. According to the Intergovernmental Panel on Climate Change (IPCC)—the group of scientists tasked by the United Nations with advising the world on climate change—warming of more than two degrees Celsius (2°C) may prove an existential threat to our civilization. Warming of more than 1.5°C will likely lead to catastrophic disruptions, countless millions displaced. Even the warming that has so far occurred is already being felt as floods, water shortages, crop failures, and many cascading impacts to our economies and politics. Limiting and potentially reversing climate change will require massive transformations of our fossil fuel–dependent energy, transportation, and industrial systems—

transformations that, while barely begun, are already changing how billions live.

How could fiction *not* be concerned about something with such vast impacts on the human condition? And indeed, we *need* fiction to turn its attention to the climate. Fiction is a form of entertainment, but it is also a way of examining the world. Stories help us derive meaning from senseless or banal events. Fictional characters, their internal motivations and experiences laid transparent on the page, help us better understand ourselves and others. Fiction is a proving ground for the language, images, and metaphors we use to make sense of life. If we are going to transform our entire civilization—or even if we are to simply go on living on an increasingly inhospitable planet—our arts and humanities need to be a part of propelling, navigating, and reckoning with that process.

But for the first few decades that climate change was in the public consciousness, popular literature seemed unable to wrap its collective head around the crisis. There were exceptions, of course. Science fiction writers like Bruce Sterling, Greg Egan, Octavia Butler, and Kim Stanley Robinson wrote about futures affected by CO2-driven global warming back in the 1990s; looking back further, J. G. Ballard imagined various flavors of extreme climate change as early as the 1960s, though not tied to our present-day understanding of our climate trajectory.

Still, in his 2017 nonfiction book *The Great Derangement*, novelist Amitav Ghosh argued that our authors—much like our politicians—had so far done little to deal with this problem that was a fundamental threat to our way of life. Ghosh's book considered why climate change seemed to be "unthinkable." His key thesis was that while literature was good at telling stories on an individualistic human scale, climate change was on another scale entirely: a collective, planetary scale. In particular, the novel—the premier form of contemporary literature—had been, Ghosh argued, finely honed as a tool for examining the amount of human life that could fit inside a house. The novel could dissect the meaning and intricacies of the relationships, objects, and history in that house. But ask the novel to look outside, to examine forces that flood cities, melt or dry out continents, and warm whole planets—well, it would be like using a magnifying glass to look at a storm, the sun, the vast and roiling sky.

Whatever the reason, literary fiction lagged behind in making sense of climate change. (Notable exceptions include Ian McEwan's *Solar* in 2010 and Barbara Kingsolver's *Flight Behavior* in 2012. However, given that the Kyoto Protocol, the first climate treaty, had been signed in 1997, I believe Ghosh's point about his peers still stands.) Much of the climate fiction written in the last couple of decades was often science fiction—which makes sense because, while we are already seeing significant impacts of climate change today, the bulk of the danger still lies in the future. Some in the science fiction community have argued that climate fiction is a superfluous term. After all, fiction about the future is the genre's traditional purview. And given that we know the future will contain some degree of climate change, shouldn't *all* science fiction be climate fiction now? Wouldn't separating the two onto different metaphorical or literal shelves merely cut science fiction readers off from the best thinking about the future of the climate? Doesn't such separation also diminish, in the eyes of the climate-concerned, thinking about the future that tackles a broad range of interconnected trends, as much popular and effective sci-fi tends to do? Aren't we, in fact, playing right into the hands of those literary snobs who are so keen to wrest this pertinent social issue out of science fiction's domain, freeing them to write about the future without stooping to tar themselves with our genre designations? And indeed: since the release of Ghosh's book, a good deal of literary-minded climate fiction has been published that eschews the tropes and traditions of sci-fi. However, this distinction is, in my opinion, largely a difference in how these cultural products are marketed to the public. No doubt the contestation of the term "cli-fi" will continue as more authors and readers take an interest in climate change.

I'm lingering on this question of genre categories because I believe they force us to ask what is unique and compelling about climate fiction. How do climate fiction and science fiction differ? The answer, I argue, is that all fiction that speculates about the future has embedded in it a theory of social change. In most science fiction, social change is driven by advancements in science and technology. As much as it is fiction about the future, it is perhaps even more so fiction about science. The average sci-fi story imagines brilliant discoveries, inventions, or technological transformations and plays out the ramifications on society.

This implied theory of social change is easy to miss because all futuristic fiction tends to get called science fiction. Occasionally someone points out this discrepancy—Ursula K. Le Guin's writing has sometimes been called "anthropological fiction," as it was usually more concerned with culture and social relations than gadgets and equations. But for the most part we haven't had much cause to interrogate the science-driven theory of social change because, for the last several hundred years, science and technology have indeed been a major driver in shaping how one generation's life unfolds differently than the preceding one's.

Of course, there are plenty of other theories about social change, both in the social sciences and in the more colloquial ways people make sense of history and the future. There is the "Great Man" theory that drives much American presidential hagiography. Marxism is a theory of social change in which the key historical driver is class struggle (though of course Marx was keenly interested in technology as well). One can conceive of society as primarily shaped by racial animus or shifting gender relations or arts movements.

In terms of crafting speculative fiction, "climate fiction" becomes a useful term because it lets us set aside traditional sci-fi ideas and pick up a new theory: that the biggest driver of social change in the coming century or more will be climate change. Cli-fi thinks that where and how most people will live in the future will be determined not by what technology gets invented but rather by how high the seas rise, how hot the planet gets, and a host of concomitant issues: flood rot, mandatory evacuation orders, the price of water, crop failures, asthma rates, whether it's too hot to go outside (Hudson 2019).

By using the "cli-fi theory of social change" to drive our speculative imaginings, we can more easily tell stories about solutions to or consequences of climate change that are social, economic, or political without necessarily being newly, flashily technological—shifting agricultural practices, for instance, or home insulation retrofits, carbon accounting, public transportation, or lifestyle changes around diet and travel. And we can get at what climate change will mean to most people: not vast engineering projects or atmospheric computer simulations, but the loss of homes to storms or collapsing market values, uncertainty in food and water prices, fraying or failing infrastructure, suffering caused by heat waves or category sixes or bomb cyclone blizzards.

The other aspect of climate fiction that makes it distinct from science fiction is the place-and-time-bound nature of the climate crisis. While science fiction often speaks to a particular political and technological moment, there is little doubt that science and technology will keep unfolding into the far future—if not here on Earth, then perhaps on the countless other worlds sci-fi imagines, in galaxies far, far away. The climate crisis, on the other hand, is a problem here on Earth that we have a limited amount of time to solve. If humanity doesn't begin making massive and swift changes to our energy systems and economy in the next ten years, keeping the planet to 1.5°C warming may slip out of reach. But years keep ticking by without significant declines in emissions.

Some commentators attribute the difficulty mobilizing climate action to the lack of narratives in popular culture that help the public understand and care about the crisis. Over the last several years many op-eds and blog posts have been written calling on fiction writers to fill this gap. Search the Internet for "can climate fiction save the planet" and you will find headlines asking this question from *The Atlantic*, *The Guardian*, *Smithsonian Magazine*, *The Conversation*, *LitHub*, *SyFy.com*, *Euronews.com*, *The Times of Israel*, and more. The idea is that stories—which can be intense and emotional and human in a way that charts and graphs and scientific papers rarely manage—might stir or convince otherwise apathetic or disbelieving segments of the public.

Climate fiction is part of a larger discourse around the power of stories and fiction to shape the human imagination and the power of imagination to shape politics and public policy. Political scientist Benedict Anderson argued in his book *Imagined Communities* (1983) that politics at any scale beyond the local depended on people being able to imagine themselves as part of an abstract entity or group, such as a nation-state. More recent work, across many disciplines, has explored the multilayered cognitive and social process of imagination and the central role in decision-making played by "imaginaries"; that is, the conceptions we have, individually or collectively, about specific institutions or historical forces we move through. In short: the visions we have of what the future will or should look like shape our political priorities, strategies, and values. Any attempt to deliberately transform the conditions of our world must necessarily involve transforming those visions as well.

In particular, stories (both traditional fiction and other styles, such as modern multimedia or oral storytelling) are increasingly seen as useful tools in driving those transformations of visions. Manjana Milkoreit—a scholar who has studied both international climate governance and climate fiction, and who was kind enough to provide an early review of this book—has explored new and compelling ways that cli-fi might influence climate politics, including by providing "subtle and complex lessons concerning the intricate relationship between climate, society, economics, and politics" (Milkoreit 2016). Certainly that was the aim of this book!

One qualitative survey of American cli-fi readers (Schneider-Mayerson 2018) has suggested that novels about climate change are unlikely to win over skeptics or deniers. However, the study found that such texts can, at least for a time, increase the *valence* of readers' climate concerns, pushing moderates and liberals to take the crisis more seriously and to believe, via a principle of repeated exposure, that negative climate outcomes are more likely.

While obviously I am keenly interested in climate fiction, I am a bit hesitant to make any "cli-fi can save the planet" sorts of arguments myself. It seems self-serving to claim that the kind of stories I write can change the course of such a huge global phenomenon. I even hedged against this line of thinking in the SSP3 story, in which Diya argues that "the decades of obsession with 'climate communication' were born out of a lack of analysis of actual power." Still, I agree with scholars like Milkoreit that climate fiction can and should be *useful*, even if I'm skeptical that the genre can save the world all on its own.

Another angle of cli-fi usefulness might be in using stories to make vivid pitches not just to the public but to individuals with power. Kim Stanley Robinson's 2020 novel *The Ministry for the Future* devotes significant energy to arguing that the world's centrals banks should take a direct and aggressive role in restructuring the economy to better fight climate change. Since central banks tend to be relatively free from democratic over-sight or direction, Robinson—much like his characters—is making an appeal directly to the technocrats who run the Federal Reserve, the European Central Bank, and similar institutions.

The Ministry for the Future is also notable as being one of the few major works of climate fiction to deal directly with the UNFCCC process, as the stories in this volume do. Robinson's novel charts a course toward sustain-ability via a new institution within the UNFCCC tasked with representing future generations—the eponymous Ministry for the Future. The ministry, unlike the current UNFCCC, acts unilaterally, using its independent budget to make investments in climate solutions, leveraging its diplomatic position to lobby central banks, and running black ops to help a movement of ecoterrorists and sab-oteurs make fossil fuel extraction and carbon-intensive enterprise too dangerous and expensive to continue.

This utopian vision aligns quite well with the one I laid out in the SSP1 story, in which the COP is made more democratic and the UN is given more teeth to enforce climate treaties. Similarly, the SSP1 "smooth transition" narrative was described as pushed along by direct action and occasional violence. Wonky fiction like *Ministry* and my SSP stories must tread a fine line between the pedagogical and the argumentative; between bringing lay readers up to speed with the debates being hashed out in the halls of power and shouting its own opinions into those halls, in the hopes that the powerful might prick up their ears.

Milkoreit (2017) points out that authors can serve as public intellectuals and use works of climate fiction as forts from which to launch wider forays into the discourse via book tours, interviews, op-eds, panel discussions, and so on—all of which can and do contribute to shifting public opinion and technocratic ideas about climate change. Books, both fiction and nonfiction, can also be used to attract a social media following, which can then provide authors with a platform for influencing discourse for years to come.

All this is an interesting position for writers to be in—to not just follow their own creative interests, nor write what sells well in the market, but to have a specific imperative to shape the culture, one with an increasingly imposing deadline.

Sustainability studies has a term that I think is useful here: "post-normal." Sustainability has emerged over the last thirty years as a discipline dedicated to studying and solving the thorny, real-world problems found in resource commons and other socio-ecological-technical systems, including the climate. (I should note that I first drafted this book as my master's thesis while studying

at Arizona State University's School of Sustainability.)
As a field of study and a scholarly discourse, sustainability overlaps with climate action discourse, but is quite distinct. While some in sustainability certainly do study and engage with the UNFCCC, most work at a smaller scale. Sustainability "practitioners" are often hired to help cities and firms become more environmentally friendly via supply chain efficiency, waste reduction, recycling, and switching to green materials and production practices. This is a very different kind of work than most of what is decided at the COP, such as setting and tracking emission reduction targets for whole nation-states. We can also think of sustainability as being concerned with a set of goals broader than the climate movement's—sometimes called planetary boundaries. These are entwined with the climate crisis, but it isn't hard to imagine that we could solve our greenhouse gas problem and still struggle with, say, overfishing or groundwater depletion.

Early in its development, sustainability was described by philosophers of science Silvio Funtowicz and Jerome Ravetz (1993) as a "post-normal science." "Normal science" refers to the process of knowledge creation described by Thomas Kuhn in his landmark book *The Structure of Scientific Revolutions* (1962): using the scientific method to do peer-reviewed research, testing theories to look for gaps that might lead us to new ways of thinking, replicating and confirming experimental results. All within a mostly values-free context of relatively low stakes and relatively clear facts.

I think this is analogous to how many fiction writers have long operated. What we create might matter to our individual careers, but the stakes for society at large are most often pretty minimal. Meanwhile questions of what constitute the "right" things to write are settled by success in the marketplace.

Post-normal science situates itself outside the realm of pure research or applied science, outside offering advice to those trying to set policy (see fig. 2). Post-normal science is science that happens in contexts where "facts are uncertain, values in dispute, stakes high, and decisions urgent" (Ravetz 1999).

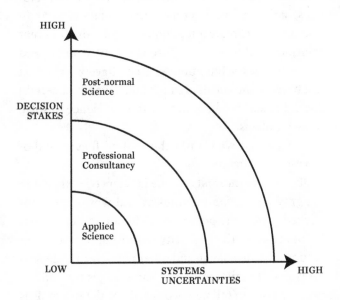

Figure 2

Post-normal science often tries to use science to engage with "wicked problems." To summarize the formulation by design theorists Horst Rittel and Melvin Webber (1973), wicked problems are problems where there is no single correct view, different views may lead to contradictory strategies, and you only get one shot. Climate change falls into this category—so much so that it has been called a "superwicked problem." Superwicked problems have all the difficulties of wicked problems, but on top of that there is no central authority, the actors trying to fix the problem are also the ones causing the problem, and time is running out (Levin et al. 2012).

While a great deal of science can be done to tell us about the dynamics of the climate or of our energy systems, in the end we don't have other copies of Earth to test different climate mitigation strategies on. Different solutions will produce different sets of winners and losers. It's messy. But Funtowicz and Ravetz argue that just because the stakes are high and facts are uncertain doesn't mean we shouldn't try to use science to solve these problems.

I argue that we should think of cli-fi as a kind of "post-normal fiction."

Just as post-normal science is science entangled in the gray-zoned complexities of real-world, one-shot wicked problems, post-normal fiction is fiction that is entangled in the precarity of our cultural moment around climate change and other narrative-influenced disasters. It is fiction that acknowledges the limited time we have to tell new stories about the future. It is fiction that tries, on some level, to motivate collective action. It is fiction that is willing to tread at the edge of propaganda and activism for the sake of writing work that meets a cultural need. It is fiction that knows that if those of us who want to see climate action do not write the stories about climate change, those who don't care are more than happy to occupy that lane. After all, one could argue that the most impactful piece of climate fiction ever written is Michael Crichton's 2004 novel *State of Fear*, whose narrative calling climate change a vast scientific hoax (backed up by misleading graphs and appendixes) was cited as evidence by climate deniers in the US Congress.

Climate fiction is not only written to encourage political action. It can also be written to explore and understand the possible futures ahead of us. This is not to say that climate fiction is or can be predictive. This is true even in the most scientifically rigorous cli-fi works, such as Robinson's *Science in the Capital* trilogy or *New York 2140*. Perhaps a better frame is to think of climate fiction as what environmental humanities scholar Joni Adamson (2013) calls a "seeing instrument."

The "seeing instrument" concept abstracts the notion of foresight to account for many different ways of examining the future. A "seeing instrument" might be a telescope, looking predictively from where we are to where we might go. But it could also be a magnifying glass, allowing us to peer around *inside* an imagined future, even if we aren't sure how we got there. It could be an astrolabe or other navigational device used to chart a future's location after we've been blown off course. Metaphors abound.

Being able to see in many ways with many kinds of instruments is useful because there is no singular climate future. Climate change is going to mean a million different things to a million different communities, a billion things to a billion individuals. It will mean something different ten years from now, fifty years from now, one hundred, two hundred, a thousand years from now. There will be bad years and better years. There is a wide-open space of possibility to explore. And all of it is still unwritten—the choices we make in the coming few decades can steer us into one climate future or another. If we want people to make *good* choices, we need to help them understand now what different lives might be lived in and beyond the climate crisis.

One challenge in asking people to push for radical changes to our civilization is the "devil you know" problem. Whatever you think of our present, status quo world, you know that it has a place for you. It might be a miserable place, but you're still here (until of course, you aren't). But there's no guarantee of place in a vague and hypothetical future, even one that will be theoretically better. So, shaking up the status quo, even a lousy one, feels risky. As sci-fi writer Karl Schroeder recently put it in a tweet, "It's easier to plan for how to personally thrive in a dystopian future that you think you understand, than a utopian one that you don't."

Scenarios are particularly useful at breaking people out of this kind of thinking by complicating simplistic normative questions about what the future holds. For instance, rather than asking whether this future is a utopia or a dystopia, futurist quadrants like the SSPs ask who the winners or losers of different futures might be, how "good" and "bad" might be constructed in those futures, and what underlying assumptions lead us into one future or another. Fiction can do this as well.

There is a power in helping readers imagine themselves in different climate futures, particularly preferable ones. Fiction can do what scientific models cannot: imagine what the future feels and smells like, what might make us happy or sad or bored, what might fill a human day with hours, memories and meaning. These imaginings aren't quite knowledge, but they do empower us to make more thoughtful choices about what kind of future we want to live in and what we need to do now to get there.

Post-normal climate fiction can be one tool for making sense and meaning out of the enormity of the climate crisis; for creating narratives that drive positive collective action on climate; and for exploring the possibility spaces of diverse climate futures.

In March 2019, in the early stages of writing this book, I attended the Scenarios Forum at the University of Denver, a conference on using the SSPs to guide modeling and communication. Many of the details about the conference are a blur, as the sessions themselves were overshadowed by the unpleasant experience of slogging through a real-life climate event: the "bomb cyclone" blizzard that hit Denver, canceled the conference early, and left me stranded for several days. A year later the stories were done, and I defended my thesis via Zoom amidst the fearful chaos of the first months of the COVID-19 pandemic. A year after that I polished up my revisions during an unprecedented heatwave that brought absurd temperatures to northern climates, all during a news week that also featured floods and hurricanes and pipeline explosions. "It's disaster season," I half-joked, grim, to a friend.

During these sorts of events, I often wonder how necessary climate fiction is, given that one can also simply turn on the news, or even look out one's window, and see the crisis unfolding in real time. More pressing: If I am really serious about doing everything I can to stop climate change, how do I know that writing is the best way to spend my time? Perhaps I would be more effective spending those hours doing political organizing, or running for office, or manually installing solar panels, or simply planting trees. Maybe I should work the highest-paying job I can get (not writing, surely) and donate every spare cent to CAN or the Sunrise Movement. Maybe I should chain myself to construction equipment to block a pipeline. Perhaps nothing I can contribute through climate fiction or any of those other activities outweighs, in the final judgement, the damage I do simply by being a living consumer. If the nature of post-normal fiction is to engage with a time-bound problem that we get only one shot to solve—if it is fiction trying to do something more than entertain—then this risk looms over every page.

But that is precisely the nature of the post-normal territory: stakes are high, answers are uncertain, values are in dispute. Even though the crisis is already here, we still have work to do and debates to hash out, regardless of which SSP or other scenario we end up in. So, much like my characters, and the COP, and humanity in general, we have little choice but to do our best to muddle through.

ACKNOWLEDGMENTS

This book is indebted to the support of my committee at Arizona State University: Paul Hirt, who advised me throughout my entire graduate student career; Sonja Klinsky, who first told me about the SSPs and who helped me get to the COP to do my research; Matt Bell, who has taught me much about the craft of fiction; and Ron Broglio, whose radical and critical perspective informs key moments throughout the book. Care About Climate helped me find a place to stay in overbooked Katowice during COP24. Bas van Ruijven and Narasimha Rao of IIASA generously met with me when I visited Vienna, and that conversation inspired the core conceit of this book. And finally, every page of this book would be different without the presence of C, my partner on this writing journey, who reads everything I write and who has helped me grow as an author in countless, countless ways.

WORKS CITED

Adamson, Joni. 2013. "Environmental Justice, Cosmopolitics, and Climate Change." In *The Cambridge Companion to Literature and the Environment*, edited by Louise Westling, 169–83. Cambridge: Cambridge University Press.

Anderson, Benedict. 1991 [1983]. *Imagined Communities: Reflections on the Origin and Spread of Nationalism*. 2nd ed. London: Verso.

Atwood, Margaret E. 2015. "It's Not Climate Change—It's Everything Change." Medium, July 27, 2015. https://medium .com/matter/it-s-not-climate-change-it-s-everything-change-8fd9aa671804.

Buckell, Tobias S. 2018. "A World to Die For." *Clarkesworld* 136 (January). http://clarkesworldmagazine.com/ buckell_01_18/.

Burgess, Matthew G., Justin Ritchie, John Shapland, and Roger Pielke Jr. 2020. "IPCC Baseline Scenarios Have Over-projected CO2 Emissions and Economic Growth." *Environmental Research Letters* 16 (1): 014016.

Crichton, Michael. 2004. *State of Fear*. New York: HarperCollins.

Funtowicz, Silvio, and Jerome Ravetz. 1993. "Science for the Post-Normal Age." *Futures* 25 (7): 739–55. https://doi.org/10.1016/0016-3287(93)90022-L.

Ghosh, Amitav. 2017. *The Great Derangement: Climate Change and the Unthinkable*. Chicago: University of Chicago Press.

Hudson, Andrew Dana. 2019. "Why Climate Fiction?" *Imaginary Papers* on Medium, January 29, 2019. https://medium.com/imaginary-papers/why-climate-fiction-967876f650fd.

IPCC. 2021. Climate Change 2021: *The Physical Science Basis. Contribution of Working Group I to the Sixth Assessment Report of the Intergovernmental Panel on Climate Change*. Edited by V. Masson-Delmotte, P. Zhai, A. Pirani, S. L. Connors, C. Péan, S. Berger, N. Caud, Y. Chen, et al. Cambridge: Cambridge University Press, in press.

Kingsolver, Barbara. 2012. *Flight Behavior*. New York: HarperCollins.

Kuhn, Thomas S. 1962. *The Structure of Scientific Revolutions*. Chicago: University of Chicago Press.

Levin, Kelly, Benjamin Cashore, Steven Bernstein, and Graeme Auld. 2012. "Overcoming the Tragedy of Super Wicked Problems: Constraining Our Future Selves to Ameliorate Global Climate Change." *Policy Sciences* 45:123–52. https://doi.org/10.1007/s11077-012-9151-0.

McEwan, Ian. 2010. *Solar*. London: Vintage Books.

Milkoreit, Manjana. 2016. "The Promise of Climate Fiction: Imagination, Storytelling and the Politics of the Future." In

Reimagining Climate Change, edited by Paul Wapner and Hilal Elver, 177–91. London: Routledge.

———. 2017. "Imaginary Politics: Climate Change and Making the Future." *Elementa: Science of the Anthropocene* 5:62. https://doi.org/10.1525/elementa.249.

Nikoleris, Alexandra, Johannes Stripple, and Paul Tenngart. 2017. "Narrating Climate Futures: Shared Socioeconomic Pathways and Literary Fiction." *Climatic Change* 143 (3): 307–19.

O'Neill, Brian C., Elmar Kriegler, Kristie L. Ebi, Eric Kemp-Benedict, Keywan Riahi, Dale S. Rothman, Bas J. van Ruijven, et al. 2017. "The Roads Ahead: Narratives for Shared Socioeconomic Pathways Describing World Futures in the 21st Century." *Global Environmental Change* 42 (January): 169–80.

Ravetz, Jerome R. 1999. "What Is Post-Normal Science." *Futures: The Journal of Forecasting, Planning and Policy* 31 (7): 647–53.

Rittel, H. W., and M. M. Webber. 1973. "2.3 Planning Problems Are Wicked." *Polity* 4:155–69.

Robinson, Kim Stanley. 1984–90. *Three Californias* trilogy: *The Wild Shore* (1984); *The Gold Coast* (1988); *Pacific Edge* (1990). New York: Tom Doherty.

———. 2004–7. *Science in the Capital* trilogy: *Forty Signs of Rain* (2004); *Fifty Degrees Below* (2005); *Sixty Days and Counting* (2007). New York: Bantam Books.

———. 2017. *New York 2140*. New York: Orbit.

———. 2020. *The Ministry for the Future*. New York: Orbit.

Schneider-Mayerson, Matthew. 2018. "The Influence of Climate Fiction: An Empirical Survey of Readers." *Environmental Humanities* 10 (2): 473–500. doi: https://doi.org/10.1215/22011919-7156848.

Thill, Scott. 2009. "Margaret Atwood, Speculative Fiction's Apocalyptic Optimist." *Wired*, October 20, 2009. https://www.wired.com/2009/10/margaret-atwood-speculative-fictions-apocalyptic-optimist/.

Andrew Dana Hudson is a speculative fiction writer and sustainability researcher. He is a fellow in the Imaginary College at the Arizona State University Center for Science and the Imagination and is a member of the 2020 class of the Clarion Workshop. His fiction won the Imagination and Climate Futures Initiative's 2016 Climate Fiction Contest and has been featured in *Slate*, *Vice*, *Lightspeed*, and more. He lives in Tempe, Arizona.